T0240099

SpringerBriefs in Energy

SpringerBriefs in Energy presents concise summaries of cutting-edge research and practical applications in all aspects of Energy. Featuring compact volumes of 50 to 125 pages, the series covers a range of content from professional to academic. Typical topics might include:

- A snapshot of a hot or emerging topic
- A contextual literature review
- A timely report of state-of-the art analytical techniques
- An in-depth case study
- A presentation of core concepts that students must understand in order to make independent contributions.

Briefs allow authors to present their ideas and readers to absorb them with minimal time investment.

Briefs will be published as part of Springer's eBook collection, with millions of users worldwide. In addition, Briefs will be available for individual print and electronic purchase. Briefs are characterized by fast, global electronic dissemination, standard publishing contracts, easy-to-use manuscript preparation and formatting guidelines, and expedited production schedules. We aim for publication 8–12 weeks after acceptance.

Both solicited and unsolicited manuscripts are considered for publication in this series. Briefs can also arise from the scale up of a planned chapter. Instead of simply contributing to an edited volume, the author gets an authored book with the space necessary to provide more data, fundamentals and background on the subject, methodology, future outlook, etc.

SpringerBriefs in Energy contains a distinct subseries focusing on Energy Analysis and edited by Charles Hall, State University of New York. Books for this subseries will emphasize quantitative accounting of energy use and availability, including the potential and limitations of new technologies in terms of energy returned on energy invested.

More information about this series at http://www.springer.com/series/8903

Meysam Qadrdan · Muditha Abeysekera ·
Jianzhong Wu · Nick Jenkins ·
Bethan Winter

The Future of Gas Networks

The Role of Gas Networks in a Low Carbon
Energy System

 Springer

Meysam Qadrdan
Cardiff University
Cardiff, UK

Muditha Abeysekera
Cardiff University
Cardiff, UK

Jianzhong WU
Cardiff University
Cardiff, UK

Nick Jenkins
Cardiff University
Cardiff, UK

Bethan Winter
Wales & West Utilities
Newport, UK

ISSN 2191-5520   ISSN 2191-5539   (electronic)
SpringerBriefs in Energy
ISBN 978-3-319-66783-6   ISBN 978-3-319-66784-3   (eBook)
https://doi.org/10.1007/978-3-319-66784-3

This Springer imprint is published by the registered company Springer Nature Switzerland AG
The registered company address is: Gewerbestrasse 11, 6330 Cham, Switzerland

Acknowledgements

The authors would like to acknowledge Mr. Lahiru Jayasuriya (Cardiff University) for contributing to preparing Chap. 2 and Mr. Oliver Lancaster (Wales and West Utilities) for providing Fig. 4.3.

Contents

Chapter 1
Overview of the Transition to a Low Carbon Energy System

European countries intend to reduce their emissions of Greenhouse Gases (GHG) up to 100% by 2050 compared to 1990 levels [1] and many other countries around the world have similar ambitions. Achieving these targets requires a substantial transformation of the energy systems in those countries that are reliant on fossil fuels to meet their energy needs. In particular, a reduction of the emissions from generating electricity and providing heat is expected to play a vital role in decarbonising energy systems worldwide.

In this chapter a brief overview of the major changes in the energy systems of European countries will be provided. The expected impacts on gas networks of decarbonising the electricity and heat sectors, as well as the potential role of gas networks in a low carbon energy system will be summarised to prepare the reader for more detailed discussions in Chaps. 3, 4 and 5.

1.1 Changes in the Gas Sector

Natural gas currently supplies around 22% the energy used worldwide. In many countries, natural gas supplies a substantial amount of the energy used for heating and power generation. Natural gas consumption is expected to increase in many regions in the world, with the exceptions of Europe and Eurasia (see Fig. 1.1). This trend of increasing global natural gas consumption is a consequence of replacing more polluting energy sources such as coal and oil with natural gas, as well as the increasing demand for energy in developing countries.

Natural gas demand in Europe increased from 490 bcm[1] in 2015 to 548 bcm in 2017 (approximately 12% increase) primarily due to demand in the power sector to

---

[1] Billion cubic meter of natural gas in standard temperature (20 °C) and pressure (1 atmosphere).

© The Author(s), under exclusive licence to Springer Nature Switzerland AG 2020
M. Qadrdan et al., *The Future of Gas Networks*,
SpringerBriefs in Energy, https://doi.org/10.1007/978-3-319-66784-3_1

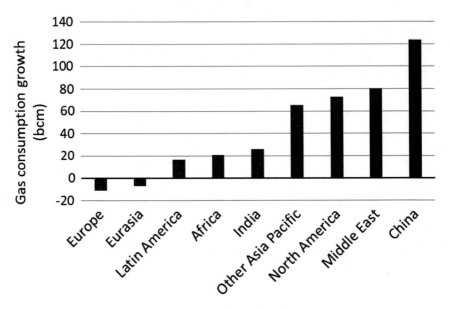

Fig. 1.1 Prediction of natural gas consumption growth from 2017 to 2023 [2]

compensate for the decommissioning of coal power stations [3]. However, according to the IEA [2] the demand for gas in Europe is expected to decrease by 11.4 bcm between 2017 and 2023. This is due to improving energy efficiency, an increasing share of renewable sources for power generation, and the move towards electrification of heat.

Gas demand for power generation will be reduced by the increase in wind and PV generation. Owing to their flexible operating characteristics, gas-fired generating units will play a crucial role in compensating for the variability of renewable energy sources. Consequently, variations of wind and solar generation will be transferred to the gas demand [4]. The wide-scale integration of variable renewable generation in power systems is expected to decrease annual gas demand for power generation, but potentially lead to larger and more frequent fluctuations in the demand for gas [5, 6].

1.2 Decarbonisation of the Power System

Many European countries have already made impressive progress in deploying renewable and low carbon energy sources for electricity generation. In 2017, the share of renewable sources in total electricity generation in Europe was almost 30%. As shown in Fig. 1.2, in Europe since 2010 the increasing use of wind and solar generation coincides with reducing output from coal fired power stations. This is a consequence of the incentives that have been introduced for promoting renewable energy, as well as regulations such as the Large Combustion Plant Directive [7] that

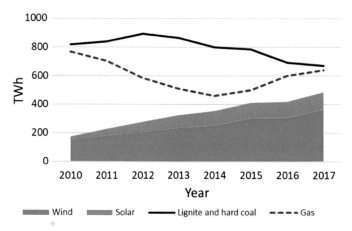

Fig. 1.2 Electricity generation in Europe. This figure has been re-produced using data published in [8]

have minimised the use of coal power stations that are not fitted with equipment to reduce emissions of SO_2 and NO_X.

On the other hand, there is an inverse correlation between the share of gas and coal in power generation. The recent increase in electricity production by gas-fired power plants has acted to compensate for the fluctuations in renewable generation whilst coal power plants are closing.

The rising fraction of electrical energy generated by gas is expected to be a short-term trend. With the increased use of renewable and low carbon generation, the role of gas-fired power plants is changing from providing constant output, base load generation with a high capacity factor to acting as peaking plant. This means that although gas-fired plants are needed at times to supply the gap between electricity demand and generation from renewable and nuclear generation, the capacity factor of gas-fired plants will be reduced, which consequently results in the reduction of annual gas demand for power generation.

1.3 Decarbonisation of the Heat Sector

In the European Union, heating and cooling accounts for almost half of the final energy consumption. In 2017, only 19.5% of heating and cooling demand was supplied by renewable sources of energy. It is evident that the decarbonisation of heat will have a significant role in decarbonising the economy.

Table 1.1 shows the fraction of different energy sources for supplying heat in Europe. Gas supplied approximately half of the heat demand.

The Baltic and Nordic countries have the highest share of renewable energy sources used in heating, ranging from 43% in Estonia to 67% in Sweden. Biomass is the most widely used source of renewable energy used for heating today,

Table 1.1 Share of different
primary energy sources in
supplying heating and cooling
in Europe in 2012 [9]

Fuel	%
Gas	46
Coal	15
Oil	10
Other	4
Biomass	11
Wind, PV, Solar, Geothermal	7
Nuclear	7

representing some 90% of all renewable heating. The role of gas in supplying heat varies for different countries across Europe. Countries in Western Europe such as UK, Germany, France, Netherlands, Italy and Belgium, currently consume significant volumes of natural gas every year to meet their heat demand in domestic and commercial buildings [10].

This book starts with providing fundamental information about gas networks (Chap. 2), and attempts to (i) provide an overview on how decarbonisation of the power and heat sectors affect gas demand and the way gas networks will be operated in future (Chaps. 3 and 4). It concludes by discussing the potential role for gas infrastructure in a low carbon energy system (Chap. 5).

References

1. EU Climate Action Available on: https://ec.europa.eu/clima/citizens/eu_en. Accessed 17 Sep 2018
2. IEA Gas 2018: analysis and forecasts to 2023. Available on: https://www.iea.org/gas2018/. Accessed 17 Sept 2018
3. Honor A Natural gas demand in Europe in 2017 and short-term expectations. Oxford Institute for Energy Studies, 2018. Available on: https://www.oxfordenergy.org/wpcms/wp-content/uploads/2018/04/Natural-gas-demand-in-Europe-in-2017-and-short-term-expectations-Insight-35.pdf
4. Oswald J, Raine M, Ashraf-Ball H (2008) Will British weather provide reliable electricity? Energy Policy 36(8):3202–3215
5. Qadrdan M, Chaudry M, Jenkins N, Baruah P, Eyre N (2015) Impact of transition to a low carbon power system on the GB gas network. Appl Energy 151:1–12
6. National Grid (2012) Gas ten year statement 2014
7. Directive 2001/80/EC of the European Parliament and of the council of 23 October 2001 on the limitation of emissions of certain pollutants into the air from large combustion plants, 2007
8. Jones D, Sakhel A, Buck M, Graichen P (2017) The European Power Sector in 2017
9. European Commission An EU strategy on heating and cooling, 2016. Available on: https://ec.europa.eu/energy/sites/ener/files/documents/1_EN_ACT_part1_v14.pdf
10. Honor A Decarbonisation of heat in Europe: implications for natural gas demand. The Oxford Institute for Energy Studies, 2018. Available on: https://www.oxfordenergy.org/wpcms/wp-content/uploads/2018/05/Decarbonisation-of-heat-in-Europe-implications-for-natural-gas-demand-NG130.pdf

Chapter 2
Fundamentals of Natural Gas Networks

2.1 Natural Gas

Natural gas is a fossil fuel that was formed over millions of years from decomposing organic matter, which was subject to intense heat and pressure as successive layers of sand and rock were laid down. It is a mixture of hydrocarbon and non-hydrocarbon gases and is usually found in porous material underneath an impervious rock layer that prevents the gas from reaching the surface and escaping into the atmosphere. Conventional natural gas deposits are commonly found in association with oil reservoirs, with the gas either mixed with the oil (associated gas) or floating on top of it (non-associated gas). The term unconventional gas is used to describe shale gas, tight gas and coalbed methane (see Chap. 5).

The principal constituent of most natural gases is methane. There may also be smaller amounts of heavier hydrocarbons including propane and butane as well as certain non-hydrocarbon gases such as carbon dioxide, nitrogen, hydrogen sulphide and helium. Natural gas that is suitable for commercial exploitation typically contains 80–95% methane and low concentrations of other gases, as shown in Table 2.1.

The physical and chemical properties of the gas mixture influence the design and performance of gas appliances and the associated gas supply infrastructure. When Britain converted from using manufactured town gas (a mixture mainly of hydrogen and carbon monoxide) to North Sea natural gas in the 1960s many gas appliances and fittings had to be modified or replaced.

The properties of natural gas depend primarily on its composition. For example, the relative density (or specific gravity), gas compressibility and the calorific value of a gas mixture are approximately the weighted averages (by molar fraction) of the individual gases. Some of the gas properties of natural gases are shown in Table 2.1.

© The Author(s), under exclusive licence to Springer Nature Switzerland AG 2020
M. Qadrdan et al., *The Future of Gas Networks*,
SpringerBriefs in Energy, https://doi.org/10.1007/978-3-319-66784-3_2

Table 2.1 Composition of natural gas from different sources. Reproduced from Table 2 in Ref. [1]

	North Sea[a]	Groningen, Netherlands	Lacq, Pyrenees[b]	Brega, Libya[c]
Methane (mol%)	93.63	81.3	97.4	66.8
Ethane (mol%)	3.25	2.85	2.2	19.4
Propane (mol%)	0.69	0.39	0.1	9.1
Butane (mol%)	0.27	0.13	0.05	3.7
C_5 and higher (mol%)	0.20	0.06		<1.0
Carbon dioxide (mol%)	0.13	1.0		–
Nitrogen (mol%)	1.78	14.3	0.3	–
Helium (mol%)	0.05	0.05		–
Type	UK, sweet dry	Dutch sweet dry high N_2	France, dry	N. African LNG (total)
Molecular weight	16.96–18.95	18.62	16.43	22.87
Relative density @15 °C, 1 atm	0.59–0.66	0.6438	0.5682	0.7927
Density @15 °C, 1 atm $\left[\text{kg/m}^3\right]$	0.71–0.80	0.789	0.696	0.971
Calorific Value (Gross) @15 °C $\left[\text{MJ/m}^3\right]$	37.97–39.2	41.36	50.78	57.3

[a] Average received, Bacton terminal
[b] Purified
[c] Composition of LNG ex Brega Liquefaction plant

The combustion of methane is an exothermic oxidation process which produces thermal energy and water (H_2O) as a by-product.

$$CH_4(g) + 2O_2(g) \rightarrow CO_2 + 2H_2O(g) + 810 \text{ kJ of thermal energy}$$

where, (g)—gas; CH_4—methane (g); O_2—oxygen (g); H_2O—water; CO_2—carbon dioxide (g)

2.2 Natural Gas Flow in Pipes

Gas flow in pipes is described as laminar or turbulent. The Reynolds number (Re), a dimensionless parameter, is defined to characterise fluid flow in pipes as,

$$Re = \frac{\rho V D}{\mu} \tag{2.1}$$

where,

ρ is the fluid density $\left[\frac{kg}{m^3}\right]$
V is the fluid velocity $\left[\frac{m}{s}\right]$
D is the pipe internal diameter [m]
μ is the viscosity $\left[\frac{N.s}{m^2}\right]$

The Reynolds number is an indication of the relative importance of inertia forces with respect to viscous forces in the fluid flow.

Laminar flow occurs at low Reynolds numbers (typically below 2000). Such a flow occurs at low gas velocities in smooth pipes. In laminar flow, gas particles move in an orderly fashion with layers of fluid appearing to slide over each other (see Fig. 2.1). The gas towards the middle of the pipe moves at a greater velocity than that close to the pipe's internal surface.

Turbulent flow occurs when laminar flow breaks down at higher gas velocities. Two types of turbulence are defined depending on whether the eddies develop only in the centre of the pipe (partial turbulence) or whether they fill the whole pipe (full turbulence).

Figure 2.1 is an illustration of laminar and fully turbulent flow in pipes. The arrows indicate the path of a fluid particle. Natural gas flow in pipeline networks is usually partially turbulent or turbulent due to the low viscosity of natural gas, high flow velocities and the effect of bends and fittings in the pipes.

For isothermal gas flow in horizontal pipelines, the general steady-state gas flow equation is shown in Eq. (2.2) and is derived from an energy balance [2]. The friction factor f in Eq. (2.2) is based on experimental results. Several formulations are proposed by different authors for estimating the friction factor. Some common formulae are shown in Table 2.2.

$$Q_n = \sqrt{\frac{\pi^2 R_{air}}{64}} \times \frac{T_n}{p_n} \times \sqrt{\frac{(p_1^2 - p_2^2)D^5}{f\,SLTZ}} \tag{2.2}$$

where,

Q_n is the volume flow rate of gas at T_n and p_n $\left[\frac{m^3}{s}\right]$

R_{air} is the specific gas constant for air $\left[287.058\frac{J}{kg\cdot K}\right]$

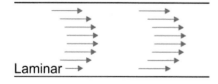
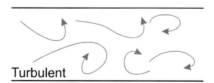

Laminar   Turbulent

Fig. 2.1 Laminar and full turbulent flow in pipes

Table 2.2 Common formulae for estimating the friction factor in medium and high-pressure gas pipes

Formula name	Application	$(1/\sqrt{f})$
Polyflo	Used in medium pressure networks operating between 0.75 and 7 bar(gauge) pressure	$5.338 \times Re^{0.076} \times E^{a}$
Rough-pipe (fully turbulent)	Characterizes most high pressure gas operating conditions	$4\log(\frac{3.7D}{k})^{b}$
Weymouth	Used in high pressure networks operating above 7 bar(gauge)	$20.64 \times D^{0.167} \times E$
Panhandle 'A'	Used in high pressure networks with large diameter transmission pipes	$6.872 \times Re^{0.073} \times E$

[a]E— pipeline efficiency factor. This is a correction factor used to represent errors in actual flow rate from that predicted by the flow equation due to extra friction effects such as from fittings and bends, weld beads, dirt and rust scaling on the internal wall of pipes
[b]k—effective roughness of the pipe (height of surface irregularities) [m]

T_n is the standard temperature [273.15 K]
p_n is the standard pressure [1 bar]
p_1, p_2 are the gauge pressures at the pipe inlet and outlet $[bar_g]$. All pressure measurements are typically expressed as gauge pressures in gas network systems. It is the amount by which the gas pressure exceeds that of the atmosphere.
D is the pipe internal diameter [m]
f is the friction factor
S is the relative density of gas
L pipe length [m]
T is the average gas temperature in the pipe [K]
Z is the compressibility factor

For low pressure networks operating between pressures of 0–75 mbar (gauge), a simpler form of the general flow equation is,

$$Q_n = 5.72 \times 10^{-4} \times \sqrt{\frac{(p_1 - p_2)D^5}{f\,SL}} \tag{2.3}$$

with $p[mbar]$, $D[mm]$, $L[m]$ then Q_n is obtained in $\left[\frac{m^3}{h}\right]$

And the value of f is given by Unwin's formula as,

$$f = 0.0044 \times \left(1 + \frac{12}{0.276}D\right) \tag{2.4}$$

2.3   Gas Networks and Pressure Management

In many countries, natural gas is transported over long distances and distributed to industrial, commercial and residential units via an interconnected pipeline network. The pipeline system connects all consumers to a single network to exploit diversity of the demand and economies of scale in gas supply.

Natural gas networks typically have a hierarchical structure with several pressure levels being used for the transport and distribution of gas. This is similar to the use of different voltage levels in electricity networks. High pressures are used for long distance transport of gas whereas low pressures are used for gas distribution closer to small consumers. The main advantage of using higher pressures is that a large quantity of natural gas can be compressed and transported using lower cost pipes (smaller diameter but with high strength). Higher operating pressures also result in lower pressure drops over long distances and reduce the need to re-compress gas.

Gas networks can be described under three headings as,

- Gas terminals or gas entry points
- Gas transmission
- Gas distribution.

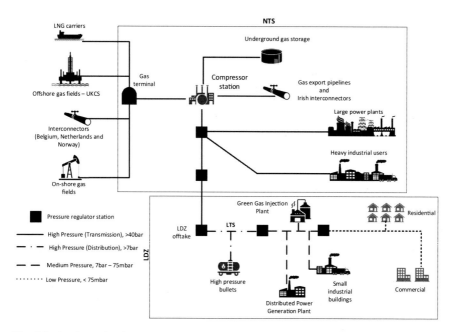

Fig. 2.2   A schematic of gas supply system in Great Britain. NTS: National Transmission System; LTS: Local Transmission System; LDZ: Local Distribution Zone; UKCS: United Kingdom Continental Shelf; LNG: Liquified Natural Gas

Figure 2.2 shows a simplified schematic of the structure of the gas network in Great Britain. It illustrates the boundaries between different systems and elements that connect to each system.

2.3.1 Gas Terminals

Gas terminals are the entry points to a network and ensure gas is within specified standards for transport and distribution. There can be a number of gas sources that connect to gas terminals such as,

- On and offshore natural gas fields
- Liquefied natural gas (LNG) imports
- Gas available from neighbouring countries through pipeline interconnectors.

Natural gas from onshore and offshore gas fields is often mixed with oil. At the gas terminals the gas and oil are separated and the contaminants such as water, carbon dioxide, hydrogen sulphide and any solids removed to produce pipeline-quality dry natural gas.

Liquified Natural Gas (LNG) terminals are purpose built ports that receive LNG from specialised cargo ships known as LNG carriers. LNG is stored at cryogenic temperatures and re-gasified before injection into the gas transmission system.

Pipeline interconnector terminals connect gas transmission systems between neighbouring countries. Interconnectors enable single or bi-directional gas exchange between countries to facilitate energy trading. The national gas transmission system (NTS) in Great Britain (GB) is connected to Belgium, Netherlands and the Republic of Ireland via pipeline interconnectors.

2.3.2 Gas Transmission Systems

The gas transmission system is operated at high pressure. In GB the national transmission system (NTS) operates at gas pressures between 38 and 94 bar (gauge). Typically, the pipes are of high-strength steel to handle the high pressure. Compressor stations boost the gas pressure that is lost through friction as the gas moves through the steel pipes. Transmission pipelines are typically buried underground; the diameter of most of the pipes used in the GB NTS range from 900 to 1200 mm.

There are some large-scale gas storage facilities, such as underground salt caverns, that are used to maintain security of gas supply. However, the variation of pressure within the transmission pipeline network itself is used as short-term gas storage to balance supply and demand variations within a day. This volume of gas at pressure within the pipeline system is called the *linepack*.

In GB, *National Grid Gas plc* owns and operates the gas transmission network.

2.3.3   Gas Distribution

When gas leaves the transmission system and enters gas distribution networks the pressure is reduced through a number of stages in regulator stations and finally delivered to consumers. Depending on the number of pressure reduction stages the gas distribution system may be further categorized. In GB, there are four distribution pressure levels: high (>7 bar(gauge)), intermediate (2–7 bar(gauge)), medium (0.075–2 bar (gauge)) and low pressure (<75 mbar(gauge)) networks. Gas pipes in the distribution system are typically of steel, PVC or cast iron construction.

2.4   Components of Gas Networks

In addition to the pipes, gas networks contain gas compressors, gas storage, pressure regulators and valves.

2.4.1   Gas Compressors

A gas compressor increases the pressure of a gas by compressing it (by doing work on the gas stream) and reducing its volume. It supports the operation of the gas network by compensating for pressure drops and storing gas in the pipelines. In the GB NTS there are 24 compressor sites with 75 compressors on 7600 km of gas transmission pipelines.

Gas compressors are of two types,

- Positive displacement type, (gas is physically trapped between two moving elements) e.g. reciprocating compressor
- Non-positive displacement type, (a rotating component imparts its kinetic energy to the gas stream) e.g. centrifugal compressor.

Both centrifugal and reciprocating compressors are widely used in natural gas networks.

Centrifugal compressors consist of three main parts; the impeller, diffusor and the volute (see Fig. 2.3a). Work is done on the gas stream by an impeller that is rotated by a prime mover. The work translates into kinetic energy and increases the velocity of gas leaving the impeller. In the diffusor the velocity of gas is reduced and the kinetic energy is converted to potential energy in the form of pressure. The gas stream then enters the volute casing where it is collected and discharged at a high pressure.

Reciprocating compressors are positive-displacement machines where the gas is trapped and compressed between a moving piston and a cylinder volume to increase its pressure (see Fig. 2.3b). A reciprocating compressor is similar in design to a

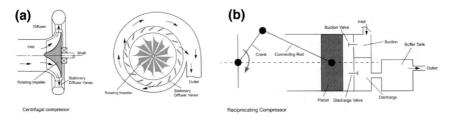

Fig. 2.3 Schematics of compressors **a** Centrifugal compressor **b** Reciprocating compressor

reciprocating engine, however its function is reversed. A prime mover is connected
through a crankshaft and connecting rods to a series of pistons with reciprocating
motion. The outward motion of the piston opens the suction valve and gas is drawn
into the cylinder from the suction manifold. Once the piston is in its outermost position the suction valve closes and the piston moves to compress the gas volume inside
the cylinder. When the gas pressure inside the cylinder volume exceeds the pressure
in the delivery manifold the delivery/discharge valve opens and the compressed gas
flows at an approximately constant pressure to the delivery manifold. The pressure
falls instantaneously when the piston starts to move on a new suction stroke, the
discharge valve closes, the suction valve opens and the cycle is repeated.

The main advantages and disadvantages of the two compressor types are summarised in Table 2.3.

Different types of prime movers are used to drive gas compressors. Gas turbines,
gas engines, steam turbines and electric motors have all been used. Economic and
environmental conditions influence the choice of prime mover.

The power required from the prime mover to provide a specified pressure ratio is
given by,

$$N = \frac{p_1 Q_{in}}{\eta (n-1)} \left[\left(\frac{p_2}{p_1} \right)^{\frac{n-1}{n}} - 1 \right] \tag{2.5}$$

Table 2.3 Advantages and disadvantages of centrifugal and reciprocating type compressors

Type	Centrifugal compressor	Reciprocating compressor
Advantages	Few moving parts Maintenance cost and lubrication oil consumption are low High capacity flow rate Continuous delivery of gas without variations	High compression ratios
Disadvantages	Low compression ratios	Lower mechanical efficiencies More moving parts

where,

N is the power required from the prime mover [W]

p_1, p_2 are the suction and discharge pressures [Pa]

Q_1 is the inlet gas flow rate $\left[m^3/s \right]$

n is the polytropic index of a gas

η is the compressor overall efficiency

    Multi-stage compressor units can be arranged in series and/or parallel to obtain the pressure ratios and handle the gas flow rates required.

2.4.2 Gas Storage

Gas storage supports real time network management and provides security of gas supplies. It offers additional gas supply capacity in different timeframes i.e. day-to-day or seasonal, by storing gas when demand is low and releasing it during high demand.

    The key technical characteristics of a gas storage system are its storage capacity (in million cubic meters or GWh), the rate of gas injection and withdrawal (in million cubic meters or GWh per hour or per day), the turnover rate and cushion gas. Turnover rate is the number of times in a year the storage facility could be filled and emptied. All storage systems require a volume of gas to be retained to deliver the required rate of gas withdrawal. This is called 'cushion gas' and represents a significant cost as it is not recovered until the storage unit terminates operations.

    Gas storage methods can be classified as low pressure holders, high pressure holders, linepack, underground gas storage and Liquefied Natural Gas (LNG) storage.

    Low pressure gas holders are large tanks erected above ground that store gas at low pressure and ambient temperature. They are typically of two types.

- Water sealed holders: An inverted open cylindrical container placed in a vessel of water, that rises and falls within a steel guide frame to make space for gas.
- Dry/waterless holders: A stationary cylindrical shell, within which a piston rises and falls to allow gas storage.

    In GB, low pressure holders were originally built during the period of manufactured gas to act as a buffer between gas generation and demand. These units have now been made redundant by advances in pressure management systems and linepack storage. In GB all low pressure gas holders have been decommissioned.

    High pressure gas holders can be classified as two types.

(a) Large spherical or cylindrical tanks: These are also termed bullets and frequently used by gas utilities for short term balancing purposes.

(b) Bottle and pipe-type underground holders: These are arrays of bottle or pipe type gas containers connected in a series and parallel configuration and buried underground. They are used particularly for small storage requirements. This type of storage has lower initial cost compared to gas storage in bullets or spheres (Fig. 2.4).

Fig. 2.4   Gas storage **a** Bullets **b** Low pressure gas holder **c** Salt cavern gas storage. *Photos* a With permission from Wales and West Utilities, UK. b Kemp, Martin/Shutterstock.com. c Avvakumova, Maria/Shutterstock.com

Linepack: When the high pressure pipeline network is used as a means of gas storage, it is termed *linepack*. Gas is stored by allowing more gas to enter a pipeline than is being withdrawn thus increasing the pressure, packing more gas in the pipe section. Linepack gas storage capacity can be increased by

(a)   increasing the operating pressure in the pipes using compressors
(b)   increasing the diameter of the pipes
(c)   paralleling a portion of a pipe with additional pipes, or a combination of these means.

Linepack storage can be implemented at relatively small capital and operating expense and is used typically for balancing gas demand with supplies within a day.

Underground storage: Natural gas is stored underground when it can be injected into natural rock or sand reservoirs that have suitable connected pore spaces. These are typically depleted oil or gas fields and salt caverns. Aquifers are also used by displacing water with gas. Essential reservoir features include,

(a)   an impermeable reservoir cap to prevent leaks and pressure loss
(b)   high porosity and permeability in the reservoir rock
(c)   the depth of the reservoir sufficient to allow a safe pressure
(d)   either an absence of water or easily controlled water conditions in the reservoir
(e)   a thick vertical reservoir formation rather than a thin horizontal formation
(f)   an oil free formation (although exhausted oil-producing formations have been used successfully).

Underground storage is a cost-effective way to provide a large capacity for seasonal gas storage.

LNG storage: Liquefied Natural Gas is used for either shipping or storing natural gas. Liquefaction reduces the volume of gas and requires much smaller volumes than other systems to store the same quantity of gas. LNG storage tanks can be located either on the surface or underground. LNG is a cryogen (a liquid that boils at very low temperatures) and is stored at very low temperatures (~ -162 °C) to maintain its liquid form. Tanks usually consist of two concentric containers where the inner tank stores LNG and the outer container is filled with a thermal insulation material. Tanks vary greatly in size, depending on the application. LNG is rapidly vaporised using a heat source and delivered to the gas network typically during contingencies such as network constraints or failures in supply.

Depending on their function, storage systems can be classified as,

- Peak shaving: Gas is injected and withdrawn at high flow rates and used to deal with high demands that last for a short period of time (e.g. LNG storage)
- System support: Storage facilities in the network are used to balance short term variations between gas supplies and demand (e.g. bullets, linepack)
- Multi cycle or flexible storage: Gas is injected and discharged several times a year (e.g. Salt caverns)
- Seasonal storage: Gas is mainly injected in summer and withdrawn in the winter (e.g. depleted oil or gas fields)
- Strategic: Gas that is stored for use in an emergency (e.g. aquifers).

Table 2.4 shows the main advantages and disadvantages of the different types of gas storage. The right combination of storage systems considering the principal advantages and economies of each type helps to provide a more reliable supply of natural gas.

2.4.3 Pressure Regulators

Pressure regulators are used to automatically control the pressure and hence the flow of gas to downstream processes and customers. They are located at interfaces between different gas pressure levels in the network and at consumer connections. An ideal pressure regulator would supply downstream gas demand while maintaining the discharge pressure constant. Low gas pressures are required for residential, commercial and some industrial uses. Before regulators came into use, an operator had to observe a pressure gauge throughout the day for pressure variations, which indicated downstream gas demand. When pressure decreased more gas flow was required. The operator would then manually open a control valve until the pressure increased showing that downstream demand was being met.

A modern pressure regulator automatically varies the rate of gas flow to maintain a specified outlet pressure. In its simplest form, it can be modelled as a variable size orifice inserted into the pipeline. The area of the orifice is a function of the difference between the outlet pressure and a desired outlet pressure.

The mass flow rate of gas through a simple orifice is given by,

$$\dot{m} = \frac{A_2 C}{B} \sqrt{2\rho_1(p_1 - p_2)} \tag{2.6}$$

where,

$\dot{m}$ is the mass flow rate through the orifice $\left[\text{kg/s}\right]$

p_1, p_2 are the inlet and discharge pressures [Pa]

ρ_1 is the density of gas at the inlet side $\left[\text{kg/m}^3\right]$

Table 2.4 Advantages and disadvantages of gas storage types

Technology		Function in the gas network	Advantages	Disadvantages
Underground	Salt caverns	Multi cycle/flexible	High injection and withdrawal rates, low cushion gas, phased development, relatively high turnover rates	Small volume in individual cavern
	Depleted oil/gas field	Seasonal/strategic	Existing and understood, low cost, large capacity	High cushion gas, slow injection and withdrawal rates, Low turnover rate
	Aquifers	Seasonal/strategic	Large capacity	High cost, extended development time, Environmental issues, Low turnover rate
High pressure holders	Bullets (Methane)	Peak shaving/system support	High injection and withdrawal rates, high turnover rates	Small volume
	Linepack	System support	Low cost	Complexity in controls
Low pressure holders	Gas holders	System support	High turnover rates	Large space and safety, environmental issues
Liquefied natural gas (LNG)		Peak shaving/system support	High withdrawal rate	High cost, Low capacity, Safety

$$B = \sqrt{(1 - (A_2/A_1)^2}$$

A_1 is the pipe area $\left[\mathrm{m}^2\right]$
A_2 is the orifice area $\left[\mathrm{m}^2\right]$
C is the outlet (discharge) coefficient defined as the ratio of the actual volumetric flow rate to the ideal value

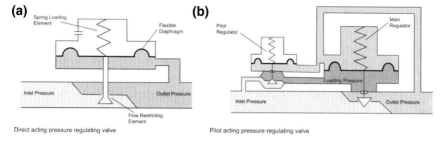

Fig. 2.5 a Direct operated valve schematic **b** Pilot operated valve schematic

Pressure regulators can be classified as,

- Directly operated
- Indirectly/relay operated or pilot-operated.

Directly operated gas regulators are units in which the controlled pressure is directly used to position the flow restriction valve. Directly operated valves have three essential elements (see Fig. 2.5a),

- A flow restricting element: This may be in various forms such as a poppet valve, disk, piston or slide valve.
- A loading element: This is used to apply a force to which the controlled pressure is compared. This force may be applied by a dead weight or a spring
- A pressure sensing element: This is usually a flexible diaphragm that moves according to the difference between the controlled pressure and the force applied by the loading element. The flow restricting element is attached to the diaphragm with a stem so that the flow opening is controlled by the motion of the diaphragm.

The controlled pressure applied to the pressure sensing element i.e. the diaphragm, acts as a valve-closing force. The valve opening force is provided by the loading element. The force imbalance between the loading element and the controlled pressure determines the position of the pressure sensing element and the flow restricting element and thereby controls the size of the orifice opening.

Ideally, pressure regulators should give a constant outlet pressure at all flows from zero up to the capacity of the valve. However, the outlet pressure may change due to,

- Variation of the spring constant over length of stroke of valve
- Inlet pressure variation
- Minimum flow
- Valve body design
- Diaphragm area change.

Direct-operated valves have limited ability to accommodate large variations of gas flow and to maintain a constant gas pressure at discharge. Pilot-operated regulators were designed to overcome these shortcomings and increase sensitivity of pressure control.

Pilot-operated gas regulators use a pilot or a pressure amplifier to increase regulator sensitivity. Analysis of pilot-operated regulators can be simplified by considering them as two independent direct-operated regulators connected together as shown in Fig. 2.5b. The smaller of the two is the pilot. The pilot valve senses the controlled pressure directly and adjusts the smaller spring-open valve to control the loading pressure on the main regulator valve. The main regulator is used to control large flows and pressure variations with higher accuracy.

2.4.4 Other Valve Types

A number of valve types are used in a gas networks to perform functions such as,

- Isolation valves: Used to completely interrupt flow and isolate a section of the pipe. Gate, plug, ball and butterfly valves are commonly used for this purpose.
- Check valves: Used to prevent reverse gas flow in pipelines.
- Pressure relief valves: Used to relieve excessive gas pressure build up in pipelines to prevent equipment damage or failure.

The basic requirement in the design of these valves is that they offer minimum resistance to flow when open. In many types of calculations it is often justified to neglect the pressure losses through such devices.

2.5 Gas Demand

Estimating and forecasting gas demand is an important task in the design, capacity planning and day-to-day control of the gas network. Gas demand forecasts are used for a variety of purposes such as,

- Ensuring security of supply: Demand forecasts are essential to ensure sufficient gas is held and can be made available to maintain gas supply without interruptions in the short, medium and long term.
- Investment planning: Demand forecasts are the main driver for planning future investment in gas network infrastructure.
- Operation planning: Demand estimates are used by gas utilities for the safe, reliable and efficient delivery of gas to consumers in real time.
- Pricing: Demand forecasts and estimates are also used for gas pricing.

Gas demand can be categorised by considering the end consumer as,

- Residential customers
- Commercial customers
- Industrial customers and
- Electrical power generation.

Residential and commercial customers: The gas demand of residential and commercial customers is driven by space heating and hot water demand. Gas demand for heating depends primarily on weather conditions and consumer behaviour (e.g. thermostat setting, hot water use). Space heating demand typically establishes peak conditions and affects the seasonal variability of gas demand. Gas network operators define a single daily measure of weather for each local distribution network termed the *Composite Weather Variable* which takes into account the wind speed, effective temperature and the seasonal normal effective temperature to characterise the average weather condition. The near-linear relationship observed between gas demand and the Composite Weather Variable makes it a useful parameter to model and forecast gas demand. Figure 2.6 shows the variation of gas demand in the South Wales local distribution zones (LDZ) in the UK plotted against the Composite Weather Variable. The deviation from the linear curve arises primarily from non-heating gas use. Non-heating gas demand (e.g. cooking) is typically estimated by applying appropriate load and diversity factors to the total connected load.

Industrial customers: The gas demand of industrial customers is difficult to generalise due to the wide variation in both size and activity among enterprises. It is typically defined in categories depending on the nature of the business. One suggested approach is to analyse the largest customers in each of the subclasses of industry (e.g. steel) in the region. The forecast is then based on historic patterns of consumption consistent with the expected level of installed capacity in each subclass of industry.

Electrical power generation: Gas demand for power generation is estimated based on the installed gas fired electricity generation capacity and on how frequently the generators are used. Electricity demand forecasts, weather forecast, relative fuel prices and the installed capacity of renewable generation are all taken into account when forecasting gas demand for power generation. The age of individual gas fired

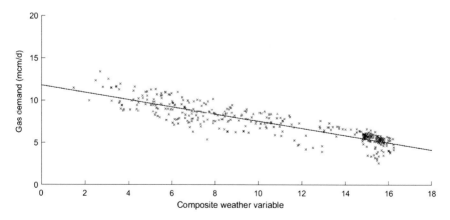

Fig. 2.6 Gas demand with respect to the Composite Weather Variable. (Produced using data for South Wales in 2016—data available on [http://www2.nationalgrid.com/data-item-explorer/])

power stations, commercial arrangements, government policies and environmental legislation are considered when forecasting which plants will be used and closed in future.

2.5.1 Methods to Define Gas Demand

The following methods are used to characterise gas demand.

- **1 in 20 peak day demand criterion** is the maximum demand for gas that will occur in a single day, in not more than 1 winter in 20 years. It represents an extremely high level of gas demand and is used in the design of gas networks and planning for extreme weather conditions.
- **Load duration curves** are used to provide an estimate of the total gas demand and its duration above a specific demand threshold (expressed as a %). They are constructed by plotting the daily gas demand in a year in order of decreasing magnitude i.e. with the greatest demand at the left and the lowest demand at the right (see Fig 2.7). Gas demand data from a series of years is used to create load duration curves for average and extreme conditions and are used for planning purposes.
- **Hourly/daily/weekly/seasonal gas demand profiles** are used to assist with short term operation of the gas network. Mathematical models based on historical demand data and weather forecasts are used to generate gas demand profiles that are used for planning the short-medium term operating strategy by optimising the use of gas storage and minimising the costs of gas delivery.

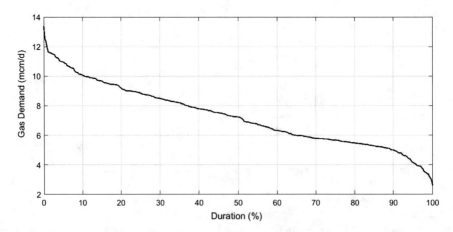

Fig. 2.7 An example curve illustrating the fraction of time in a year gas demand exceeds a given value (Produced using data for South Wales in 2016—data available on http://www2.nationalgrid.com/data-item-explorer/)

2.6 Operation and Control of Gas Networks

The main responsibility of a gas network operator is to maintain safe and efficient operation of the gas supply system in real time and to ensure security of supplies in the short, medium and long term.

The deployment of gas stocks and storage and the control of gas pressure and flows in a network are typically managed at different geographical scales. In GB, the national gas transmission network is owned and operated by National Grid. National Grid is responsible for delivering gas of the right quality, quantity and pressure at each of the local distribution zone offtake points and direct connected industrial sites. The regional gas distribution networks are owned by 4 different companies and controlled by control centres in each sub-region. The gas distribution network operators are responsible for extracting gas from the transmission network and delivering it to the end consumers.

Day to day control of the gas transmission and distribution system requires metered data to be transmitted from the points of measurement to the control room via telemetry systems. The following parameters are regularly monitored to assist in gas network control:

- Gas flow rate at entry and exit points in the network, at selected locations in the network and into and out of storage units.
- Gas pressure at compressor station inlet and outlet, pressure reduction stations and at selected key points in the network.
- Gas stocks and storage levels.

Metered data and gas quality information is processed automatically and the information required for control displayed in a control room. Control room staff regularly check that the gas stocks and the gas flows are satisfactory to meet the forecast demand in the forthcoming period and maintain linepack and gas storage levels as required. A number of optimisation algorithms, implemented in computer programs, are used to assist the planning and optimal control of the operation of the gas network. Control room staff ensure that gas pressure is maintained at the agreed levels at key points in the network. Critical pressure points in the network are monitored regularly. Manual adjustments for gas flows and pressure can be made as required by changing the operating points in compressor stations and control valves. The actual and predicted performance of the system is regularly monitored and a frequently revised forecast of the profile of gas input, demand and stock levels maintained for the next period.

Figure 2.8 shows a typical 24 h operation of the gas distribution network in the South Wales region, UK, illustrating the rate of gas input, gas demand and variation of networks linepack. Linepack is maintained approximately at the same level at the end of each 24 h cycle to ensure that gas demand can be met in case of sudden peaks and supply disruptions.

Control room staff are also responsible for network condition monitoring and automatic signals are typically built into the software systems to indicate a condition

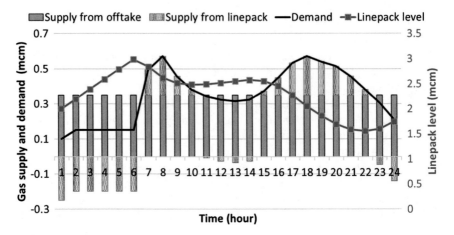

Fig. 2.8 Variation of gas demand, gas supply from offtake and linepack, gas stock levels in a 24 h operation of a gas distribution network [3]

of alarm. Planned and unplanned maintenance activities are carried out by maintenance teams in co-ordination with the control room staff to minimise any supply disruptions.

References

1. Lom WL, Williams AF (1976) Substitute natural gas: manufacture and properties. Wiley, New York
2. Osiadacz AJ (1987) Simulation and analysis of gas networks. J.W. Arrowsmith Ltd, Bristol
3. FREEDOM project final report, October 2018. Accessed on 21/06/2019 on https://www. westernpower.co.uk/projects/freedom

Further Reading

4. Gas engineers handbook. American Gas Association, Industrial Press Inc, New York, 1965
5. Gas transmission and distibution. D. L. Copp, Walter King Ltd.. London, 1967
6. Gas making and natural gas. British Petroleum, Ben Johnson & Co. Ltd, 1972
7. Gas Demand Forecasting Methodology, NationalGrid, 2016

Chapter 3
The Operation of Gas Networks in the Presence of a Large Capacity of Wind Generation

The growing use of renewable energy for generating electricity in those countries that have relied on gas-fired power stations to meet the bulk of their electricity demand is affecting gas consumption. In this Chapter, the impacts of a substantial capacity of wind generation on gas demand and consequently on the operation of gas networks are discussed. In addition, the potential role of the gas infrastructure as a source of flexibility to future low carbon power system is described. The gas and electricity systems of Great Britain (GB) were used as a case study.

3.1 Case Study: GB Gas and Electricity Systems

3.1.1 The Electricity System in GB

3.1.1.1 Current and Future Electricity Demand

Great Britain (England, Scotland and Wales) has a single interconnected electrical power system with Direct Current (DC) links to France, the Netherlands, Northern Ireland and the Republic of Ireland. It has an electricity peak demand of around 62 GW which according to the future energy scenarios produced by National Grid[1] [1] is anticipated to increase in the long term under a range of scenarios that have been developed to explore future national energy needs: Two Degrees, Slow Progression, Steady State and Customer Power [2] (Fig. 3.1).

---

[1] National Grid operates gas and electricity transmission systems in GB

[2] **Two Degrees**: Two Degrees has the highest level of prosperity. Increased investment ensures the delivery of high levels of low carbon energy. Consumers make conscious choices to be greener and can afford technology to support it. With highly effective policy interventions in place, this is the only scenario where all UK carbon reduction targets are achieved.

 Slow Progression: In Slow Progression low economic growth and affordability compete with the desire to become greener and decrease carbon emissions. With limited money available, the

© The Author(s), under exclusive licence to Springer Nature Switzerland AG 2020
M. Qadrdan et al., *The Future of Gas Networks*,
SpringerBriefs in Energy, https://doi.org/10.1007/978-3-319-66784-3_3

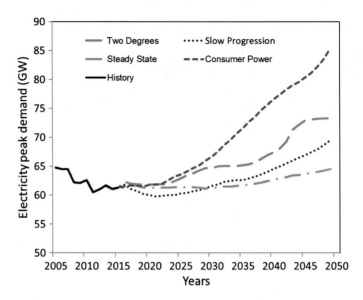

Fig. 3.1 Historical and projected electricity peak demand for GB [1]

3.1.1.2   Current and Future Electricity Supply for GB

GB in 2016 had a generating capacity of around 100 GW. According to all of National Grid's energy scenarios, wind generation will constitute a major fraction of the total generation capacity in the near future (Fig. 3.2). The anticipated capacity of gas-fired generation plants in 2025 varies between 21 and 32 GW across different scenarios, which indicates its important role to support the integration of renewable generation. However, in 2050 the role of gas-fired generation plants is more uncertain as their capacity will reduce significantly to 3.2 GW in a Two Degrees scenario, while in the Slow Progression scenario it will increase to 31.8 GW. One of the reasons for reduced capacity of gas-fired plants in the Two Degrees scenario is the uptake of electrical energy storage and power generating plants equipped with Carbon Capture and Storage (CCS), which will provide the operational flexibility needed to compensate for wind variations. In the Consumer Power scenario, it is assumed that some CCGT power plants will be replaced with small scale gas engines connected to gas distribution networks.

---

focus is on cost efficient longer-term environmental policies. Effective policy intervention leads to a mixture of renewable and low carbon technologies and high levels of distributed generation.

Steady State: In Steady State business as usual prevails and the focus is on ensuring security of supply at a low cost for consumers. This is the least affluent of the scenarios and the least green. There is little money or appetite for investing in long-term low carbon technologies.

Customer Power: In a Consumer Power world there is high economic growth and more money available to spend. Consumers have little inclination to become environmentally friendly. Their behaviour and appetite for the latest gadgets is what drives innovation and technological advancements. Market-led investments mean spending is focused on sources of smaller generation that produce short-to medium-term financial returns.

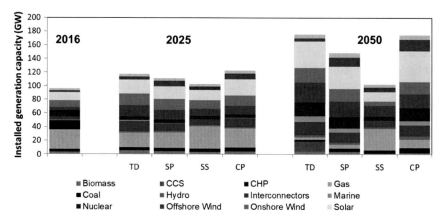

Fig. 3.2   Existing and future power generation mix for GB [1]. Columns under years 2025 and 2050 represent range of future scenarios; TD: Two Degrees, SP: Slow Progression, SS: Steady State and CP: Consumer Power. CHP stands for Combined Heat and Power

3.1.2   The Gas Transmission System in GB

3.1.2.1   Current and Future Gas Demand for GB

The annual gas demand in GB in 2016 was almost 80 billion cubic metres (bcm). Improvements in energy efficiency combined with partial electrification of the heat sector[3] in GB could decrease the gas demand to as low as 50 bcm by 2050.

Although the annual gas demand is decreasing, the peak for gas demand is likely to remain unchanged, at least until 2030, because of the reliance on gas-fired power plants to generate electricity during cold winter days with low wind speeds.

3.1.2.2   Current and Future Gas Supply for GB

The gas supply in GB comprises indigenous production in the UK Continental Shelf (UKCS), imports by pipeline from Norway and continental Europe, and imports in the form of Liquefied Natural Gas (LNG) from the global market. Production from the UKCS declined from 95 bcm in 2000 to 35 bcm in 2016. This resulted in increased gas imports to fill the gap between demand and indigenous supply. Gas imported in 2016 supplied 55% of the total demand, and this is expected to increase to 85% by 2050, due to the continued depletion of the UKCS gas reserves [1]. The share of 'green' gases that includes biogas, renewable hydrogen and other types of combustible gases from renewable sources is also expected to increase (see Chap. 5).

---

[3]In the year 2016, more than 70% of heat demand in domestic buildings in GB was met by burning natural gas in small boilers.

3.1.2.3   GB National Gas Transmission System (NTS)

Gas terminals: Gas from various sources is delivered to coastal reception terminals, and then transported to different regions across GB via a network of high pressure pipelines and gas compressors. A list of gas terminals and their supply capability is provided by Table 3.1.

High pressure gas pipeline networks: The NTS consists of 7600 km pipelines, presently operated at pressures between 38 bar up to 94 bar. Customers including storage sites, power stations and large industries are directly connected to the NTS, while local distribution networks consisting of a number of tiers of reducing pressure are used to take gas from the NTS and deliver it to small consumers in each region. A simplified representation of the GB gas transmission network and the gas terminals is shown in Fig. 3.3.

Gas storage: Gas storage facilities in the NTS support seasonal and short term balancing of gas supply and demand. A list of existing gas storage facilities and their characteristics is provided in Table 3.2. Medium range storage typically have very fast rates of injection and withdrawal that allow these facilities to go through a complete cycle of gas injection and withdrawal as market prices and demand dictate. Long range storage refers to facilities which mainly put gas into storage in the summer and take gas out of storage in the winter.

Table 3.1 Gas supply terminals in GB (IUK is an interconnector pipeline between GB and Belgium; BBL is an import pipeline between GB and Netherland; Langeled is an import pipeline from Norway) [2]

Terminal	Supply capability (mcm/day)
Bacton including IUK and BBL	151
Barrow	8
Easington including Rough storage and Langeled	118
Isle of Grain	59
Milford Haven	86
Point of Ayr (Burton Point)	2
St Fergus	102
Teesside	24
Theddlethorpe	7

Fig. 3.3 A representation of the GB gas transmission system [3] (the triangles in the map are gas supply terminals)

Table 3.2 Gas storage facilities in GB [3]

Storage facilities	Storage type	Capacity (bcm)	Maximum delivery (mcm/day)
Rough (closed in June 2017)	Depleted field	3.3	41
Aldborough	Salt cavern	0.3	40
Hatfield moor	Depleted field	0.07	1.8
Holehouse farm	Salt cavern	0.02	5
Holford	Salt cavern	0.2	22
Hornsea	Salt cavern	0.3	18
Humbly grove	Depleted field	0.3	7
Hill top farm	Salt cavern	0.05	12
Stublach	Salt cavern	0.2	15

3.2 Impacts of Variable Wind Generation on the Operation of Gas Networks

3.2.1 Wind Generation Variability

The large scale integration of wind generation into a power system poses several challenges associated with matching power supply and demand. This is because the wind is an intermittent source of energy and therefore the power generated by wind farms is variable. The relationship between electric power and wind speed for a single typical turbine, and a wind farm are shown in Fig. 3.4.

A typical wind turbine starts generating electricity at wind speeds of around 4 m/s. Between 4 and 12 m/s the power output rapidly increases to the rated generating capacity of the turbine, and remains constant for wind speeds between 12 and 25 m/s. For wind speeds above 25 m/s the turbine is shut down to avoid mechanical damage. Typically, after shut down a wind turbine restarts after 3 min of wind speed being below 20 m/s.

As can be observed from the wind turbine power curve of Fig. 3.5, at wind speeds of 4–12 m/s and also around 25 m/s, small variations in wind speed result in large changes in the power output. With many turbines connected in a wind farm, the aggregate power output of the farm is smoother than the output from a single turbine. Although, the aggregate power generation from wind farms dispersed across a wide geographical region like a country is less sensitive to the average wind speed in any region, it still shows a significant degree of hourly variation.

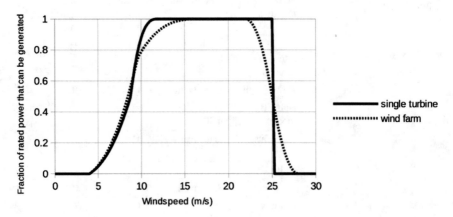

Fig. 3.4 Power curve for a single wind turbine and for a wind farm

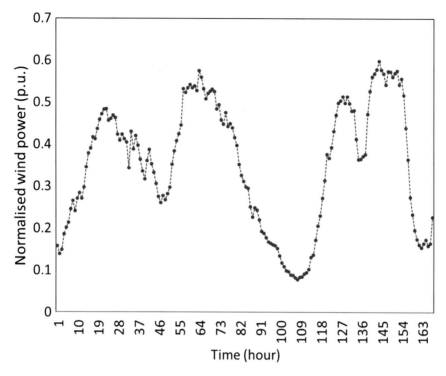

Fig. 3.5   Hourly wind power generation in GB during 10th to 16th of January 2015, normalised by total installed wind generating capacity

3.2.2   *Impacts of Wind Generation on the Net Electricity Demand*

Wind farms with their low variable operating cost and zero emissions during operation are usually given priority access to the grid and supply power whenever the wind speeds are suitable. Therefore, the variability of wind power requires other generators to continuously ramp their power output up and down in order to fill the gap between wind power supply and electricity demand. Here we define the difference between electricity demand and wind generation as *net electricity demand*. Due to ramping constraints, rapid electric power swings caused by the wind cannot be compensated by base load generation plants such as nuclear and large coal fired power stations. Hydro and pump storage plants are capable of rapid ramping but in those countries with only a small capacity of hydro-electric power (for example in GB where the capacity of hydro power stations is around 3.7 GW which accounted for only 4% of the generation capacity in 2015), gas-fired generating plants are used to compensate for the wind variability. Although gas-fired generating plants, normally Combined Cycle Gas Turbines (CCGTs), can have flexible operating characteristics, their variable operation leads to large changes in the gas demand as the CCGTs ramp up and down to fill the gap between wind generation and electricity demand.

To demonstrate how the increased use of wind generation in a power system affects the net electricity demand, real operational data for Great Britain was used.

3.2.2.1   Hourly Real Wind Generation

Figure 3.5 shows hourly real historical wind power generation in GB between 10th and 16th of January 2015 (inclusive), normalised by the total installed wind generation capacity, which was 12 GW. The power generated from the wind varied between 10 and 60% of the installed capacity of wind power within the week, with significant and abrupt increases and reductions.

3.2.2.2   Modelled Hourly Net Electricity Demand

The national hourly electricity demand for GB during 10th to 16th of January 2015, is shown in Fig. 3.6 as the black solid line. The impacts of wind generation on

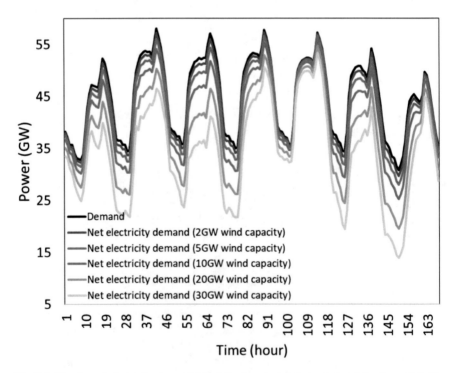

Fig. 3.6 Hourly real electricity demand (black line), and modelled net electricity demand profile for various levels of wind generation capacity

the net electricity demand was calculated assuming different levels of wind generation capacity. The hourly wind generation for different levels of wind capacity was
derived by multiplying the wind capacity by the normalised hourly wind generation.
Wind generation was then subtracted from the gross electricity demand to derive net
electricity demand. The net electricity demands are shown in Fig. 3.7 by solid grey
lines with different shades for various capacities of wind power.

The total gross electricity demand varies between 30 and 57 GW (a 27 GW swing).
Assuming the electricity demand remains the same, installing 30 GW wind turbines
in GB will result in the net electricity demand varying between 13 and 55 GW (a
42 GW swing) in this specific week.

In addition, the maximum ramp up in the gross electricity demand is 20 GW over
6 h. The maximum drop in the gross electricity demand is 25 GW over 10 h. The
maximum drop in the net electricity demand is 32 GW over 10 h, in the case of
30 GW installed wind generation capacity.

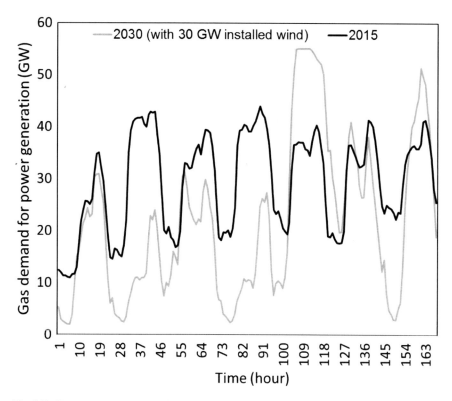

Fig. 3.7 Gas consumption by gas-fired power plants for two cases: a generation mix in 2015 with
12 GW installed wind generation capacity (black line), and a generation mix in 2030 with 30 GW
installed wind generation capacity (grey line)

3.2.3   Gas Demand for Power Generation

In power systems that use gas-fired power plants to balance electricity supply and demand, the variations in net electricity demand are transferred to the gas demand and this can have a significant impact on the operation of the gas network. The gas demand for power generation was calculated by:

$$Energy\ from\ Natural\ Gas\ =\ Heat\ Rate\ \times\ Electrical\ Energy\ Output$$

The Heat Rate (MJ/kWh) is a measure of a power plant's efficiency and is the ratio of the input energy (MJ) to the electrical energy generated (kWh).

A combined gas and electricity networks model (CGEN [4]) was used to calculate the gas demand for power generation during the week for which normalised wind generation and total electricity demand are shown by Figs. 3.5 and 3.6. The CGEN model is an optimisation tool that minimises the operational costs of gas and electricity systems simultaneously subject to meeting gas and electricity demand within operational constraints. The gas demand for power generation is calculated within CGEN based on the least cost operational strategy for the interdependent gas and electricity systems.

Using net electricity demand shown in Fig. 3.6, the CGEN model was used to calculate gas demand for power generation for a case in 2015 with 12 GW of installed wind generation capacity (shown by the black line in Fig. 3.7), and for a case in 2030 with 30 GW installed wind generation capacity (shown by the grey line in Fig. 3.7). The significant increase in wind power generation capacity leads to large changes in the gas demand for gas-fired power generation as the generators ramp up and down frequently to compensate for wind variations.

3.3   Gas Networks as Flexibility Provider

The transition to a low carbon energy system will affect the way gas networks are operated. The increasing need for flexibility to support the operation of a low-carbon power system is an important opportunity for gas networks to continue playing a crucial role in future energy systems. The abundant inherent flexibility that exists in gas networks in the form of seasonal and daily gas storage, including linepack, enables gas networks to play a role as providers of flexibility to power systems. This flexibility can be provided in different forms, including making gas available to gas-fired power station and being considered as a back-up fuel for heating in an electrified heat sector (e.g. in hybrid heat pumps).

3.3.1 Linepack

Unlike electricity, gas takes time to travel from sources of supply to demand centres. Linepack is the amount of pressurised gas within pipelines of the gas network and is used as a form of diurnal gas storage to deal with rapid changes of the gas demand and supply. Sending more gas into a pipe than is withdrawn at the downstream node results in the accumulation of gas within the pipe and increases the average pressure. Removal of more gas at the downstream node than is injected into the pipe depletes the gas within the pipe and lowers the pressure (unpacks the line).

The growing variability of gas demand for power generation, caused by wind power intermittency, will increase the variations in linepack. Operational data from National Grid (Fig. 3.8) shows that the within-day linepack swing (i.e. the difference between maximum and minimum linepack in each day) of the GB high pressure gas transmission network in 2012 fluctuated with larger magnitude (32 mcm[4]) compared to 2002 (20 mcm). The increase in the variation of linepack was due primarily to the increased capacity of wind generation and also partly as a result of the closure of gas holders in the gas distribution networks [5]. The closure of gas holders in the distribution networks reduced the gas storage capability of these networks, and required more linepack in the high pressure networks to support hourly balancing of gas supply and demand.

The linepack within the national transmission system (NTS) and local transmission system (LTS) can be used to damp the impacts of the hourly fluctuations in the gas demand on the upstream gas supply from gas terminals and large gas storage

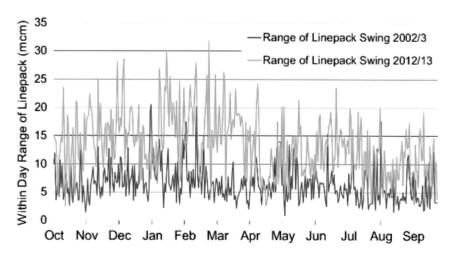

Fig. 3.8 Comparison of within-day fluctuations of NTS linepack in 2002 (blue) with 2012 (green) [5]

---

[4]1 mcm (million cubic meter) of natural gas in standard temperature and pressure contains almost 11 GWh of energy

facilities. However, the gas network operator needs to ensure that there is enough linepack within the network when an abrupt increase in gas demand is expected. Currently, National Grid Gas, balances the linepack every 24 h and makes sure that the linepack at the end of a gas day (a gas day in GB starts at 5:00 am and ends by 5:00 am of the following day) is almost equal to the linepack at the start of the gas day. The expected increase in the fluctuation of gas demand which consequently affects linepack will necessitate more dynamic (e.g. within day) linepack balancing.

The level of usable linepack is restricted by the maximum and minimum operating pressure of the pipeline system. Other fast cycle and distributed gas storage facilities (discussed in Chap. 2) can also contribute to system balancing when linepack is inadequate.

3.3.2   Gas Storage

There are different types of gas storage which play different roles in the operation of gas networks from seasonal to daily balancing. High pressure over-ground storage which also are known as *High pressure bullets* are small gas storage that each of them can store around 11 GWh natural gas at up to 30–40 bar. These storage facilities are located mostly in gas distribution networks, and their primary function is diurnal balancing of gas supply and demand at distribution level.

The growing increase in the within-day fluctuation of gas demand for power generation is expected to be addressed via a combination of linepack in high pressure gas networks and small scale and fast cycle storage in distribution networks.

3.3.3   Gas Compressors

Gas compressor stations located across high pressure transmission networks lift the pressure of gas and maintain gas flow from supply terminals to demand centres. By lifting the gas pressure, gas compressors contribute to increasing the network linepack and therefore play an important role in addressing the variability in gas demand.

The changes in the volume of gas supplied by different terminals in GB, i.e. reduction in gas supply from North Sea and increase in gas supply in the form of LNG from the south of GB, have resulted in changes to compressor utilisation. Some of the compressors are now required to support network flows in a direction reversed from their original design; some compressors have become increasingly important across a large range of demands; and some only at peak demand conditions or during certain supply patterns in order to avoid constraints. The growing reliance on imported LNG exposes the GB gas supply system to the volatile global LNG market. The variability of gas supply from LNG terminals necessitates improvement of the compressor stations to enable directing gas to different directions.

The Large Combustion Plant (LCP) Directive and the Integrated Pollution Prevention & Control (IPPC) are part of the Industrial Emission Directive (IED) which is the main EU instrument regulating pollutant emissions from industrial installations. To comply with these directives [6], the majority of gas compressor units that are driven from gas turbines in the GB gas transmission network need to be replaced by electrically driven or smaller size compressors.

Figure 3.9 demonstrates how the operating envelop of a compressor station can be affected by emission regulations. Imposing an upper limit on the emissions from a gas compressor unit constrains its feasible operating envelope in terms of pressure boost and gas throughput. In Fig. 3.10, the shaded grey area in the right hand side shows combinations of pressure boost and gas throughput that are not feasible as

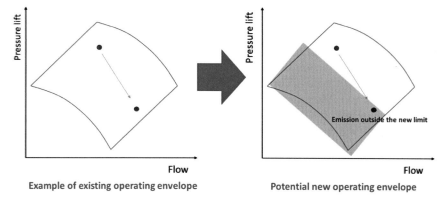

Fig. 3.9 The impact of imposing an upper limit on the emission from a compressor station on its operating envelop. Red dots are two operating points (i.e. combination of pressure lift and gas throughput). Figure reproduced from [6]

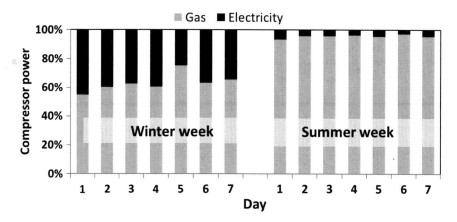

Fig. 3.10 Share of gas- and electricity-driven compressors in various days in typical winter and summer days

the emissions from the compressor unit violates the upper limit. Therefore, detailed studies are required to ensure that the reduction in the emissions from the compressor fleet will not compromise the flexibility of the gas network.

Employing electrically-driven compressors in the gas transmission network can provide flexibility to the power system by being looked upon as 'flexible electrical load'. Using the CGEN model for optimising the operation of combined gas and electricity in GB for a future low-carbon scenario [4], it was demonstrated that having a mix of electrically-driven and gas-turbine driven compressors across the gas transmission network, provides the opportunity to contribute to electricity demand balancing through switching between gas and electricity to drive compressors. Figure 3.10 shows that electrically-driven compressors operate more during a typical winter week, compared to a typical summer week. This is due to (a) higher gas demand which requires all compressor units (including electrically driven compressors) to operate with higher capacity factor, and (b) higher levels of electricity generated from wind, which potentially would be curtailed if not used by electrically-driven compressors. Comparing different days within the winter week shows a smaller share from electrically-driven compressors in day 5 which is because low wind generation caused a reduced electricity capacity margin. Therefore, the use of electricity for compressors is kept to minimum to avoid a higher electricity demand. Consequently, other gas compressors in the network compensate the reduction in the contribution of electrically-driven compressors. In the summer week, the capacity factor of wind generation is lower compared to the winter week, and at the same time, the gas demand is lower too, therefore the gas-driven compressor units are dominant.

References

1. National Grid (2017) Future energy scenarios. http://fes.nationalgrid.com/media/1253/final-fes-2017-updated-interactive-pdf-44-amended.pdf
2. National Grid (2017) Gas ten year statement. https://www.nationalgridgas.com/document/101766/download
3. Department for Business Energy & Industrial Strategy (2017) Digest of UK energy statistics. https://assets.publishing.service.gov.uk/government/uploads/system/uploads/attachment_data/file/643414/DUKES_2017.pdf
4. Qadrdan M, Ameli H, Strbac G, Jenkins N (2017) Efficacy of options to address balancing challenges: integrated gas and electricity perspectives. Appl Energy 190
5. National Grid (2014) Gas ten year statement. https://www.nationalgridgas.com/document/68796/download
6. IED Investments: Proposals Consultation 2015. https://www.nationalgridgas.com/sites/gas/files/documents/IED_Investments_Proposals_Consultation__15.pdf

Chapter 4
Impacts of Heat Decarbonisation on Gas Networks

The decarbonisation of heat is a key reason for the future decrease of the consumption of gas in European countries. In this chapter, Great Britain is used as an example to demonstrate how decarbonisation of a gas dominated heat sector could affect the demand for natural gas. Partial electrification of heat through large scale use of heat pumps is a widely discussed decarbonisation option in GB. Therefore, in this chapter the focus is on heat pump technologies and understanding how their large scale roll out could affect the operation of gas networks.

4.1 Heat Supply in GB

Heating and hot water in buildings make up around 40% of energy consumption in GB, and in year 2016 were responsible for 20% of total greenhouse gas emissions [1]. Demand for heat is met predominantly by natural gas. The large share of natural gas for heating is due to the low price of gas and the convenience of using gas boilers compared to other options. Around 22 million households in GB (84%) are connected to gas networks and use individual gas boilers to provide space heating (Fig. 4.1). Electricity has the second largest share in providing space heating. Almost 2.2 million households (8%) in GB use electricity for space heating.

To achieve the emission targets in 2050, it is crucial to find a solution for decarbonising the heat sector. The UK's Committee on Climate Change (CCC) suggests that average emissions from domestic and commercial heat should be reduced from around 220 gCO_2eq/kWh_{th} in 2015 to 180 gCO_2eq/kWh in 2030 and close to zero by 2050 [3]. A large body of research and several studies have discussed the electrification of heat as a strategy to decarbonise it. However, the electrification of heat is expected to introduce new challenges such as a sizable increase in electricity peak demand.

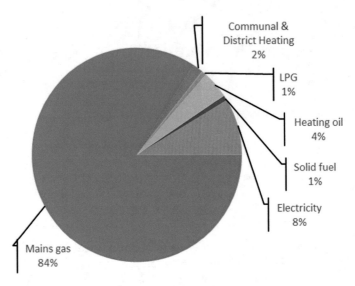

Fig. 4.1 Proportion of GB households by main space heating fuels [2]

4.2 Heat Supply Options

A range of heat supply pathways for UK have been proposed. A characteristic shared by the possible decarbonisation options is the reduced use of natural gas boilers for heating. To compensate for this reduction, alternative heating technologies such as heat pumps have been considered in these pathways.

4.2.1 Heating Technologies

Gas boilers are currently the predominant technology for producing heat in GB and many West European countries that have widespread gas networks. However, to reduce emissions and enhance the utilisation of renewable energy in the heat sector, alternative technologies and sources of energy are required. Table 4.1 shows a number of low carbon heating technologies as well as their input fuel and energy conversation efficiency.

The performance of the heating technologies in terms of their emissions depends on the emission intensity of the fuel they use and their thermal efficiency. The emission intensity of grid electricity is a measure of the level of emissions that are produced on average for generating a unit of electricity. The emission intensity of grid electricity determines whether the electrified heat is low carbon or not. For example, the heat produced by an ASHP that is powered by electricity from fossil fuels cannot be considered as low carbon heating.

Table 4.1 Selected low carbon heating technologies

Heat technology	Fuel	Thermal efficiency
Air Source Heat Pump (ASHP)	Electricity	120–400%
Ground Source Heat Pump (GSHP)	Electricity	150–500%
Direct electric heater	Electricity	100%
CHP[a] with district heating network	Gas, Biomass	45%
Micro CHP[a]	Gas	70%
Solar thermal	Solar energy	NA

[a]CHP and Micro CHP also can consume hydrogen or other renewable gases such as biogas

ASHP: Air source heat pumps use electricity and absorb heat from the outside air to supply heating to residential and commercial buildings. The operation of an ASHP is sensitive to the outside temperature and heat load. The ratio of heat generation to electricity consumption of a heat pump is known as Coefficient of Performance (CoP)—this is an indication of how efficient a heat pump is. The CoP drops as outside temperature decreases. Therefore, the lower, the outside temperature, the higher the electricity consumption by an ASHP. Performance of an ASHP is shown in Fig. 4.2. During a cold winter when the outside temperature drops below 0 °C and demand for heating is high, a supplementary source of heating (normally a gas boiler) is required to complement an ASHP.

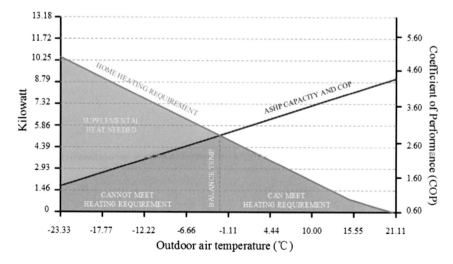

Fig. 4.2 Performance of a typical ASHP during heating seasons (adopted from [4])

Although, the heat that ASHPs extract from the outside air is renewable, the electricity needed to run the ASHPs may have environmental impact when generated. Therefore, to ensure heat produced by ASHPs is sustainable, low carbon electricity must be used to run them.

Compared to gas and oil boilers, heat pumps deliver heat at lower temperatures over longer periods. During the winter they may need to be on constantly to heat buildings effectively.

GSHP: Ground source heat pumps use a network of pipes that are buried in the ground (called the ground loop) to extract heat from the ground, and heat buildings via radiators, underfloor or through warm air heating systems. Similar to ASHPs, GSHPs deliver heat at low temperatures over extended periods.

4.2.2 Hybrid Heat Pump

If buildings are not well insulated, during cold weather when the heat load is high, an additional source of heating (usually a gas boiler) is required to complement an ASHP. The combination of an ASHP and a gas boiler is known as *hybrid heat pump*. A schematic of a hybrid heat pump is shown by Fig. 4.3. Coordinated operation of the ASHP and the gas boiler not only ensures the heat load can be met at all times, even during very cold weather, but also the fuel switching capability allows operating costs to be reduced (i.e. by switching to the ASHP when electricity is cheap, otherwise using gas). By using low carbon grid electricity when it is available, low carbon heat is produced.

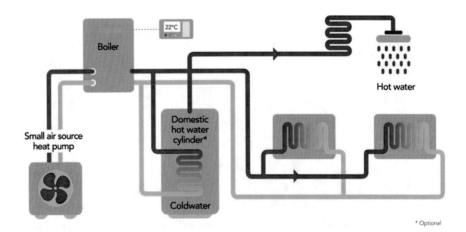

Fig. 4.3 A schematic of a hybrid heat pump. The figure provided by Wales and West Utilities [5]

4.3 Impacts of Heat Decarbonisation on Gas Networks

4.3.1 An Integrated Approach for Analysing Whole System Impacts of Heat Decarbonisation

The extent to which heat decarbonisation could affect the gas sector significantly depends on the future heating technologies. In addition, various parts of the gas networks (i.e. high pressure transmission, intermediate, medium and low pressure distribution networks) could be affected differently.

To analyse the role of gas networks in a future low carbon energy system Qadrdan et al. [7] developed a model for investigating the optimal operation of integrated gas, electricity and heat supplies in GB in 2030. The power generation mix in 2030 is shown in Table 4.2, and was based on the Gone Green scenario of Ref. [8]. For heat supply options in 2030, two extreme pathways were defined (Table 4.3) based

Table 4.2 Power generation capacity in different case studies [8]

	Power generation capacity (MW)
Nuclear	11,300
Coal + CCS	2000
Gas CCGT	27,600
Gas + CCS	1900
Wind	48,000
Solar	23,300
Hydro	1000
Interconnector	17,700
Conventional other	6800
Renewable other	8400

Table 4.3 Share of different technologies in supplying heat in 2030

	EH	CC
DHN + Gas CHP	23%	7%
Gas Micro CHP	0	5%
Gas boiler	30%	60%
ASHP	25%	10%
GSHP	5%	0
Direct electric heating	12%	15%
Oil boiler	5%	3%

These values are derived based on Customer Choice (CC) and Electrification and Heat Networks (EH) proposed by [6], however it is assumed that combined heat and power units (CHP) are fuelled by natural gas

on *Customer Choice*[1] *(CC)* and *Electrification and Heat Networks*[2] *(EH)* scenarios proposed by Ref. [6].

4.3.2   Impacts of the Heat Pathways on Electricity Demand

It was assumed that an ASHP will operate in conjunction with a gas boiler as a hybrid heating system in dwellings. This allows the option of switching between gas and electricity for meeting domestic heat demand. From the costumers' perspective, the rationale behind switching between gas and electricity in a hybrid heat pump is primarily to minimise the energy bill subject to meeting the heat demand. From the electricity system's point of view, hybrid heat pumps can be operated to avoid an increase in the electricity peak demand and its associated system reinforcements costs. For instance, when electricity peak demand is about to exceed the existing peak value, the hybrid heating systems switch to gas to supply the heat demand. However, Qadrdan et al. [7] used the performance diagram of an ASHP (Fig. 4.2) to determine the share of electricity consumption in an ASHP and gas consumption by a gas boiler to supplement the ASHP.

The electricity load duration curve in 2030 for the two heat decarbonisation pathways are shown by Fig. 4.4. These results assume that in domestic buildings that use an ASHP, a condensing gas boiler is used as a supplementary source of heating during low temperature periods.

The analysis showed that if ASHPs were to meet the heating demand on their own, the peak and annual electricity demand increases as shown in Table 4.4. The employment of hybrid heat pumps does not significantly reduce the annual electricity demand (0.2 TWh in *CC* and 0.7 TWh in *EH*). This is due to the small number of hours at which the temperature drops below 0 °C and gas boilers supplement ASHPs (see Fig. 4.2). The employment of hybrid heat pumps reduces the peak electricity demand by 600 MW in CC and 1500 MW in EH. For this analysis temperature data for 2010 was used but if the temperature drops significantly below 0 °C, the electricity peak reduction by the employment of hybrid heat pumps can be much larger.

---

[1] *Customer Choice* pathway represents a future in which the share of technologies for supplying heat demand was determined based on economically rational consumers. Consumers choose their heating system, based on upfront and running costs of different technologies and their physical fit. Gas boilers remains the predominant heating technology and will be used in 19 million homes, based on their low capital and running costs and excellent fit with UK homes. No policy intervention is assumed to meet carbon reduction targets for the residential sector.

[2] *Electrification and Heat Networks* pathway, a large contribution of district heating networks as well as a high level of electrification of heat sector through heat pumps were assumed. It was assumed by 2050 all domestic buildings use either electric heating (heat pumps and direct electric) or heat networks, fed-by zero carbon heat.

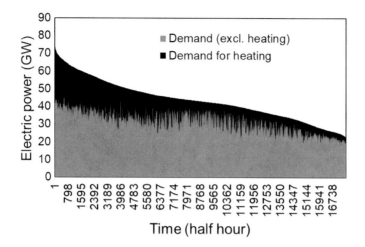

(a) Customer Choice (CC)

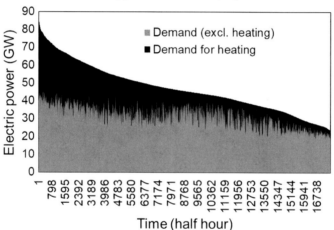

(b) Electrification of Heat and Networks (EH)

Fig. 4.4   GB electricity load duration curve in 2030 for different heat supply options [7]

Table 4.4   The impacts of hybrid heat pumps (ASHP and gas boilers) on peak and annual electricity demand

Case study	Variants	Peak electricity demand (GW)	Annual electricity demand (TWh)
Scenario CC	ASHP only	77.4	382.5
	Hybrid heat pump	76.8	382.3
Scenario EH	ASHP only	89.2	412.5
	Hybrid heat pump	87.7	411.8

4.3.3   Impacts of the Heat Pathways on Gas Demand

Gas supply duration curve for high pressure transmission network and low pressure gas distribution networks (below 75 millibar) are shown in Fig. 4.5. Compared to 2010, the annual gas supply through the low pressure networks will be reduced in 2030 by 92 TWh in *CC* and by 319 TWh in *EH*. This large reduction in the annual gas flow through the low pressure networks is primarily because of the substantial reductions in the use of domestic gas boilers to meet heat demand. The annual gas

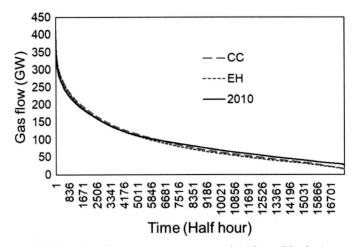

(a) High and medium pressure gas networks (above 75 mbar)

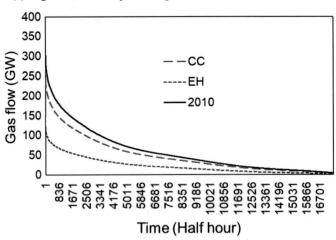

(b) Low pressure gas distribution network (below 75 mbar)

Fig. 4.5 Load duration curves for gas demand on **a** High and medium gas pressure networks (above 75 mbar), and **b** low pressure gas distribution networks (below 75 mbar)

flow in the high pressure transmission network also reduces in 2030, however the changes are not as large as seen in the low pressure networks: 59 TWh reduction in *CC* and 190 TWh reduction in *EH* (Fig. 4.5b). The smaller reductions of the gas flow in the high pressure network compared to the low pressure networks is a result of an increase in gas demand for CHP units that are connected at above 75 millibar. If it was assumed that CHP units use biomass instead of natural gas (this is an assumption made by [6]) then the shape of the gas supply duration curves for the high pressure transmission network were similar to those of low pressure distribution networks.

Electrification of heat supply and moving towards greater use of district heating networks (fuelled by natural gas) was shown to have major impacts on the low pressure gas networks. While the substitution of domestic gas boilers with heat pumps and large CHP plants (which are assumed to be connected to gas distribution networks of greater than 75 millibar) significantly reduces the maximum gas flow in the low pressure networks, the maximum gas flow in the high/medium pressure networks is expected to increase slightly compared to the reference case in 2010. This is partly due to the reliance on gas-fired power generation plants to contribute to meeting peak electricity demand, and partly due to contribution of gas-fired CHP plants in supplying heat.

The impacts of using hybrid heat pumps instead of only ASHPs on the peak and annual demand of gas were quantified for the CC and EH pathways (see Table 4.5). Both annual and peak gas demand are reduced in all the cases in which a hybrid heating system was employed. The reduction in the annual and peak gas demand happened despite the larger share of gas in supplying heat. This is due to the increased use of gas for power generation in cases in which ASHPs are used solely. During peak hours when the temperature is low and wind generation is minimal, gas-fired generators operate to their maximum capacity to supply electricity demand for ASHPs. The low CoP of ASHP during cold weather results in lower overall efficiency of heat production when gas is used in gas-fired plants and then the electricity generated is used to run ASHPs to produce heat, compared to using gas directly in a gas boiler to supply the same amount of heat. For instance, the efficiency of heat production through gas $\rightarrow$ CCGT $\rightarrow$ ASHP during cold weather is around 70% which is lower than the 90% efficiency of a gas boiler.

Table 4.5 The impacts of hybrid heat pumps (ASHP and gas boilers) on peak and annual gas demand

Case study	Variants	Peak gas demand (GW)	Annual gas demand (TWh)
Scenario CC	ASHP only	420	830
	Hybrid heating system	408	821
Scenario EH	ASHP only	422	791
	Hybrid heating system	412	788

4.4   Impacts of Heat Decarbonisation on Gas Supply Options

The changes in gas load duration curve caused by decarbonisation of the heat sector (i.e. reduced annual gas demand, but similar gas peak demand) is expected to affect the cost-effectiveness of gas supply options. In Fig. 4.6 *screening curves* superimposed on load duration curves are used to demonstrate how the changes in the gas load duration curve affects the optimal gas supply capacity.

Indigenous natural gas in addition to biogas will be used at their maximum capacity and considered to supply gas base load. The decision of how much gas to be imported via import pipeline and LNG, depends on their long term economic competitiveness.

The reduction of annual gas demand while the peak gas demand remains almost unchanged, will lead to increased supply capacity from LNG which has lower capital cost compared to import pipelines but higher gas prices. These characteristics make LNG a more economic option to meet peak gas demand.

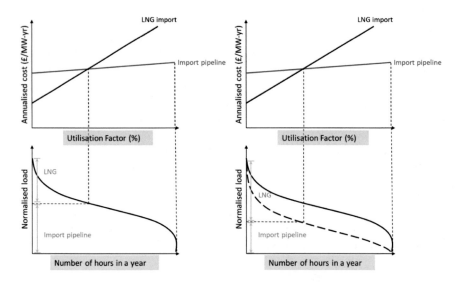

Fig. 4.6 A schematic describing the optimal mix of gas import capacity in a gas-dominated heating system (left) versus a heating system in which gas will be primarily used to meet heat peak demand (right). The dashed line on the bottom right figure presents gas load duration curve for a heating system in which gas will be primarily used to meet heat peak demand

References

1. Committee on Climate Change, Next steps for UK heat policy, 2016. Available at: https://www.theccc.org.uk/wp-content/uploads/2016/10/Next-steps-for-UK-heat-policy-Committee-on-Climate-Change-October-2016.pdf. Accessed on 23 Jun 2019
2. Ofgem, Insights paper on households with electric and other non-gas heating, 2015. Available at: https://www.ofgem.gov.uk/ofgem-publications/98027/insightspaperonhouseholdswithelectricandothernon-gasheatingpdf. Accessed on 23 Jun 2019
3. Houses of Parliament, Carbon footprint of heat generation, 2016. Available at: http://researchbriefings.files.parliament.uk/documents/POST-PN-0523/POST-PN-0523.pdf. Accessed on 23 Jun 2019
4. Overbey D Gauging the seasonal efficiency of air-source heat pumps. Walls & Ceilings online. Available at: https://www.wconline.com/blogs/14-walls-ceilings-blog/post/89498-gauging-the-seasonal-efficiency-of-air-source-heat-pumps. Accessed on 23 Jun 2019
5. Energy Management Magazine (2017) Available at: https://www.energymanagermagazine.co.uk/passivsystems-completes-first-installation-phase-for-freedom-future-of-energy-hybrid-heating-project-targeting-1-3bn-annual-savings-by-2030/. Accessed on 23 Jun 2019
6. Delta Energy & Environment, 2050 Pathways for domestic heat, 2012. Available at: http://www.energynetworks.org/assets/files/gas/futures/Delta-ee_ENA%20Final%20Report%20OCT.pdf.pdf, Accessed on 23 Jun 2019
7. Qadrdan M, Fazeli R, Jenkins N Strbac G, Sansom R (2019) Gas and electricity supply implications of decarbonising heat sector in GB. Energy 169:50–60
8. National Grid, Future Energy Scenarios, 2015. Available at: http://fes.nationalgrid.com/media/1295/2015-fes.pdf, Accessed on 23 Jun 2019

Chapter 5
The Future of Gas Networks

5.1 The Evolution of Gas Supply in the UK

Gas has been used as a source of energy since the late 18th century. The earliest discovery of flammable gas in the UK was by Thomas Shirley when he became aware of a flammable gas emanating from a spring and published his findings in the Philosophical Transactions of the Royal Society in 1667 [1]. Later, Dr. John Clayton recorded 'that water from a spring would burn like brandy' in a paper published in 1739. In 1727, Dr. Stephen Hales discovered that about one-third of the weight of Newcastle coal was distilled off in the form of flammable gases and condensable vapours when heated in a closed vessel [1, 2].

The earliest practical application of fuel gas was for lighting as a replacement for candles and oil. The Scottish engineer and inventor William Murdoch is the first to have lit a domestic residence using gas in 1792. He lit his house in Redruth, Cornwall using a gas distilled from coal. His contemporary Philip Lebon in France is known to have staged a demonstration of gas lighting in Paris at a similar time [2].

5.1.1 Gas Works

The world's first public gas utility was the London-based 'Gas Light and Coke Company' founded by Frederik Winsor in 1812 [2]. It operated the first gas works in the UK which provided street lighting using gas produced from heating coal.

A gas works also known as manufactured gas plant in the UK is an industrial plant that produces fuel gas typically from coal. Early gas works were often located beside a river or canal so that coal could be delivered by barge.

Figure 5.1 shows the key components of a typical gas works.

M. Qadrdan et al., *The Future of Gas Networks*,
SpringerBriefs in Energy, https://doi.org/10.1007/978-3-319-66784-3_5

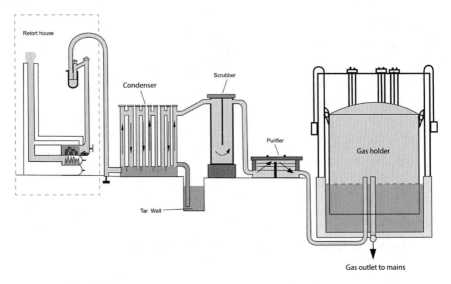

Fig. 5.1 Schematic of the key components in a gas works

A gas works consisted of several key components such as,

- A retort house—The unit where coal was heated to generate the fuel gas.
- Condenser—The unit used to remove coal tar (a dark brown or black viscous liquid of hydrocarbons and free carbon) and ammonia liquor (a concentrated solution of ammonia, ammonium compounds, and sulphur compounds) obtained as a by-product of producing coal gas by condensing it as the gas was cooled.
- Tar well tank—An underground tank used to collect the coal tar and ammonia liquor from gas condensation.
- Exhauster—A pump that was used to increase the gas pressure.
- Scrubber—The unit used to further remove ammonia and ammonia compounds from coal gas.
- Purifier—The unit used to remove hydrogen cyanide and hydrogen sulphide from the coal gas.
- Gas holder—The large vessel used for the storage of the gas and to maintain an even pressure for delivery to the distribution pipes.

The process of separating volatile gases from solid coal is called coal carbonization. The fuel gas produced is termed coal gas. It contains a mixture of gases including hydrogen (~50% by volume), methane (~35% by volume), carbon monoxide (~10% by volume) and other volatile hydrocarbons (e.g. ethane) together with small quantities of non-combustible gases such as carbon dioxide and nitrogen. The by-products from the process include coke, coal tar, ammonia and sulphur. These by-products made the business of operating a gas works very profitable.

By 1849, local gas works had been installed in over 700 large towns and cities in the UK.

The applications of gas quickly expanded beyond lighting to provide energy for heating and cooking. By 1950 about 80% of British dwellings were connected to a coal gas supply from a local gas works. The UK gas industry was nationalised in 1949 enabling the exchange of gas between neighbouring gas systems.

Gas works and gas holders were decommissioned in the UK following the introduction of natural gas in the 1970s. Coal gas is no longer produced or used in the UK.

5.1.2 Discovery and Conversion to Natural Gas

The discovery of natural gas in the North Sea led to its adoption by Britain's gas industry replacing the use of coal gas and local gas works. The main advantages of North Sea natural gas were,

- It produces less emissions thereby improving air quality
- It has a high calorific value
- It contains no carbon monoxide which is a dangerous gas that caused fatalities
- It increased security of gas and energy supplies for the UK.

North Sea natural gas is predominantly methane (~90% by volume) and has very different combustion properties to coal gas. It was necessary to adapt or replace all gas appliances (around 20 million) in a major conversion process, which started in 1967 and took ten years to complete. Although this was a large and expensive project it had the benefit of checking the safety of all gas appliances in the country and significantly reduced the number of deaths caused by gas inhalation and burns.

5.1.3 Liquefied Natural Gas (LNG)

In 1964, the UK and France made the first LNG trade, buying natural gas from Algeria. Liquefied Natural Gas (LNG) is natural gas that has been converted to liquid form for ease and safety of storage and transport. LNG is an efficient way to transport natural gas across the sea where laying pipelines is not technically feasible or economic. LNG is transported in specially designed cryogenic ships and road tankers. It is principally used for transporting natural gas to markets, where it is regasified and distributed as pipeline natural gas.

LNG takes up only about 1/600th of the volume of natural gas in the gaseous state (at Standard Temperature and Pressure). The liquefaction process involves the removal of dust, acid gases, water, helium and heavy hydrocarbons. Natural gas is then condensed into liquid form by cooling it to approximately $-162\ °C$.

The global trade in LNG is growing rapidly. The UK operates two LNG terminals, at Milford Haven in South Wales and at the Isle of Grain. 13% of the natural gas used in the UK was imported in the form of LNG in 2017.

5.2   Unconventional Natural Gas

Unconventional natural gas refers to sources of natural gas that are at present difficult to extract. It will become increasingly important as conventional gas supplies become exhausted and the cost of extraction increases. However the extractional of unconventional gas is often opposed by environmentalists and not permitted in some European countries. Unconventional gas is found in the following formations (see Fig. 5.2).

- Shale gas—Natural gas trapped in shale formations
- Tight gas—Natural gas trapped in reservoir rocks (sandstones and carbonate formations) with very low permeability
- Coal bed methane—Methane adsorbed in coal beds
- Methane hydrates (clathrates)—A solid compound found under sediments on the ocean floors in the arctic and in large undersea deposits.

Shale gas, tight gas and coal bed methane are potential sources of gas in the UK.

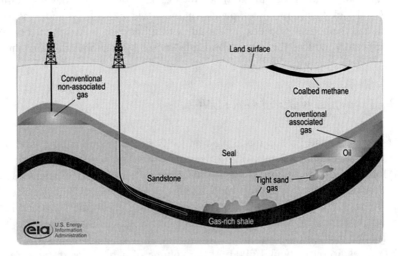

Fig. 5.2 Schematic of geology of natural gas resources. Redrawn from US Energy Information Authority web site: http://www.eia.gov/

5.2.1 Shale Gas

Shale gas is natural gas trapped within shale formations (see Fig. 5.3). Shale gas deposits are distributed over large areas on all continents. However, only the United States and Canada have exploited this resource on a large scale. Formations of shale typically have insufficient permeability to allow significant natural gas flow to a well bore. Therefore, the production of shale gas in commercial quantities requires fracturing the shale to increase permeability. The process of forming extensive artificial fractures in shale formations around gas wells is known as hydraulic fracturing (commonly called 'fracking').

Hydraulic fracturing is the injection of a fracking fluid at high pressure through a well bore into the shale formations to create cracks. The fracking fluid varies depending on the type of fracturing required and the conditions of the specific shale formation. It is typically water (90% by weight) containing sand (~9.5% by weight) and chemical additives (0.5% by weight). When the hydraulic pressure is removed from the well, the granular material in the fracking fluid (either sand or aluminium oxide) holds the fractures open. This allows the trapped gases to flow more freely into the well bore and so to the surface. In shale gas extraction horizontal or directional drilling is typically used to exploit a larger reservoir from a single well site.

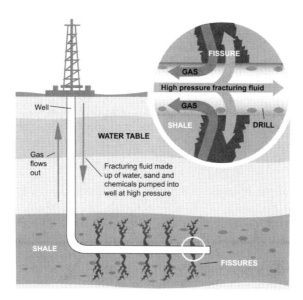

Fig. 5.3 Illustration of a shale gas extraction well with hydraulic fracturing. Redrawn from US Energy Information Authority web site: http://www.eia.gov/

5.2.2 Tight Gas

Tight gas is natural gas trapped in rock with extremely low permeability, typically sandstone or limestone. Tight gas formations were originally conventional natural gas reservoirs that were changed by cementation and recrystallization of sediments over time leading to reduced permeability of the rock and natural gas being trapped tightly within rock formations.

Tight gas extraction requires significant hydraulic fracturing to extract gas at economic rates. In addition, tight gas extraction requires deliquification or pumping water out of the well to reduce the pressure in the well and allow more gas to flow to the well bore.

Presently, tight gas extraction is carried out mainly in the United States and Canada.

5.2.3 Coal Bed Methane

Coal bed methane (CBM) is methane stored in the coal matrix by adsorption. Adsorption is the adhesion of atoms, ions or molecules from a gas or liquid to a surface. Methane in CBM is in a near liquid state, lining the pores within the coal. The open fractures in the coal (called the cleats) can also contain natural gas.

Unlike natural gas, coal bed methane contains very little heavy hydrocarbons such as propane and butane and no natural gas condensate or hydrogen sulphides. It often contains a small fraction of carbon dioxide.

To recover the coal bed methane, a bore hole is drilled into the coal layer which is typically 100–1500 m below ground level. Then the pressure within the coal bed is reduced by the extraction of natural gas or by pumping water out from the coal bed. The reduced pressure within the coal bed releases the adsorbed methane from within the coal matrix. Hydraulic fracturing may be used to release more gas from coal beds.

At present, coal bed methane extraction is practiced in the United States, Australia and Canada.

The extraction of unconventional gas typically requires hydraulic fracturing which is highly controversial in many countries. Opponents to hydraulic fracturing argue that the economic benefits of accessing gas resources are outweighed by the potential environmental impacts, which include,

- The chemicals in the fracking liquid can be dangerous and any release of the fluid could result in contamination of ground water for drinking or habitats of wildlife
- Acidization—the use of hydrochloric acid to release unconventional gas could harm workers and pollute ground water
- Drilling and hydraulic fracturing requires large amounts of water that could affect the availability of water for other uses or affect aquatic habitat

- Drilling and hydraulic fracture could release methane into the atmosphere which is a potent green house gas.

Drilling and hydraulic fracture could cause air and noise pollution and the triggering of earthquakes along with hazards to public health and the environment.

5.3 Low Carbon Substitutes for Natural Gas (Green Gas)

The global drive towards a low carbon and sustainable energy supply system has led to an increase in research and development of low carbon alternatives to fossil based natural gas in the gas grids.

The main low carbon substitutes for natural gas can be identified as,

- Biogas from anaerobic digestion processes
- Bio-Synthetic Natural Gas (Bio-SNG): Thermochemical gasification of biomass to produce synthetic natural gas
- Hydrogen derived from low carbon routes.

5.3.1 Biogas

Biogas is a mixture of gases produced by the breakdown of organic matter in the absence of oxygen. Biogas can be produced from raw materials such as agricultural waste, farm waste, municipal waste, plant material, sewage or food waste. It is considered to be a low carbon gas source because its production and utilisation is a closed loop carbon cycle.

Biogas production is a proven technology and has been widely applied globally for decades in the treatment of sewage sludge and farm waste. It is produced by the biochemical process of anaerobic digestion where naturally occurring microorganisms (methanogens) digest the organic matter and produce biogas and a nutrient rich digestate that is used as a fertilizer.

The composition of biogas varies depending on the feedstock (substrate), as well as the conditions within the anaerobic digester. Typically, biogas comprises methane (50–75% by volume), carbon dioxide (25–50% by volume), nitrogen (0–10% by volume), hydrogen (0–1% by volume), hydrogen sulphide (0.1–0.5% by volume) and oxygen (0–0.5% by volume). The calorific value of biogas is between 17 and 25 MJ/Nm^3 as compared to ~38 MJ/Nm^3 for natural gas. Biogas can be used for heating, lighting, cooking or as a fuel in internal combustion engines to generate electricity and heat.

In many European countries biogas is upgraded to match the quality of natural gas for injection into the gas network. Initially, harmful gases such as hydrogen sulphide are removed. Then carbon dioxide and other trace elements are removed to separate the methane. The most common methods used are water scrubbing and membrane technologies. In water scrubbing the carbon dioxide and other trace elements are removed by water columns running counter flow to the gas. Membrane technologies use very fine membranes to separate carbon dioxide from methane based on the size of the gaseous molecules. This purifying process can deliver near pure methane (98% by volume) and produces biomethane at this point. A final stage for conditioning the biomethane to match the gas quality criteria (e.g. calorific value, Wobbe index matching etc.) of the individual country may be required. In the UK, and other European countries such as Germany, biomethane injection in the gas grid is offered financial incentives.

Box 1: Five Fords Waste Water Treatment and Biomethane Grid Injection Plant—Dwr Cymru

Five Fords is a waste water treatment plant in north-east Wales owned and operated by Dwr Cymru (Welsh Water) a drinking water and waste water services company. The plant serves a population of around 100,000 in the neighbouring region. It is the first operational 'biomethane to grid' plant in Wales.

The plant includes two 4000 m^3 anaerobic digester tanks (see Fig. 5.2) designed to process up to 12,000 dry tonnes of sewage sludge per year. Anaerobic digestion breaks down the organic matter and produce biogas and a nutrient rich digestate that is used as a high value fertilizer on local farmland. The biogas is collected in a large tank of 1250 m^3 capacity.

Part of the Biogas is combusted in two 600 kW$_e$ combined heat and power engines to produce electricity and heat for the anaerobic digesters and waste water treatment process.

Biogas to be exported to the natural gas grid is first passed through activated carbon filters to remove hydrogen sulphide. The upgrade of biogas to biomethane uses membrane technologies. Biogas is compressed to 16 bar(gauge) and passed through the membrane plant. The very fine membranes separates methane and carbon dioxide. The carbon dioxide is vented and methane is transported to the grid injection facility for the final biomethane upgrade process.

The calorific value and other critical parameters of the gas mixture are monitored at the biomethane grid injection facility. Propane is used to enrich the biomethane (increase its calorific value) to achieve the grid compliant gas quality characteristics. Once biomethane achieves the quality standard required by Wales and West Utilities (the local gas network operator) the valve between the site and the local gas network opens and biomethane is exported to the local gas network (Fig. 5.4).

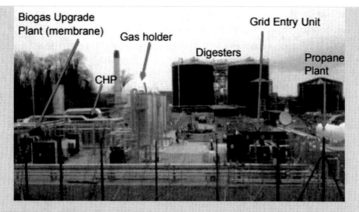

Fig. 5.4 An image of the Five Fords anaerobic digestion and gas treatment facilities. With permission from Dwr Welsh Water-Five Fords Operating Manual

For more information on the Five Fords Waste Water Treatment plant visit http://www.waterprojectsonline.com/case_studies/2016/DCWW_Five_Fords_Energy_Park_2016.pdf.

5.3.2 Bio-Synthetic Natural Gas

Bio-Synthetic Natural Gas (Bio-SNG) is produced by gasification of biomass materials (e.g. forestry residues, energy crops). Biomass gasification is a thermo-chemical process that converts organic material into combustible gases by heating in the absence of oxygen or with only limited oxygen/air.

The process of producing bio-SNG through gasification can be understood in several steps.

1. **Heating and drying**: of the biomass to produce dry organic matter while producing steam. The steam is used in subsequent chemical reactions, notably the water-gas reaction.
2. **The pyrolysis (or devolatilization)**: process occurs at around 200–300 °C where volatile gases are released from the biomass to produce porous carbonaceous char, volatile gases and tarry vapours. The process depends on the properties of the feedstock biomass which determine the structure and composition of the volatile gases, tar and char. The gas mixture consists primarily of hydrogen, carbon monoxide and often some carbon dioxide and methane. Pyrolysis can convert up to 80% (by mass) of the biomass into vapours and gases.
3. **Gas cooling and cleaning**: The gas and vapour mixture from the pyrolysis process is cooled so that contaminants and tars can be removed through thermal or physical treatment.

4. **Gas reactions with char**: The volatile gases and the char react with oxygen and steam to produce additional carbon monoxide, hydrogen and methane.
5. **Water-gas shift reaction**: The level of hydrogen in the gas mixture is boosted in a catalytic reactor by reacting carbon monoxide with steam.
6. **Methanation**: Methanation occurs in catalytic reactors and transforms hydrogen, carbon monoxide and carbon dioxide from the water-gas shift reaction into methane.
7. **Refining**: The gas at this stage contains methane, carbon dioxide, small amounts of unreacted hydrogen and elemental nitrogen. The refining stage separates these gases into the methane product gas and a carbon dioxide by-product.

The technology of biomass gasification is still the subject of research and development with a number of demonstration plants in service and some manufacturers offering equipment. Considerable heat energy is required for sustaining the chemical reaction of the gasification process. Bio-SNG holds great promise as a substitute for natural gas due to its,

- ability to use a wide range of feedstocks including mixed household waste
- conversion of over 60% of the chemical energy in the feedstock to the energy in the gas produced.

A bio-SNG plant using municipal refuse is being demonstrated in Swindon, UK [3].

Box2: Go Green Gas: Bio-SNG Production Pilot Plant in Swindon, UK

Cadent (previously National Grid Gas Distribution) partnered with Advanced Plasma Power, Progressive Energy and Schmack Carbotech to demonstrate the technical and commercial viability of Bio-SNG production from household waste. The partners demonstrated a pilot plant that produce Bio-SNG from waste collected at households in Swindon, UK (Fig. 5.5).

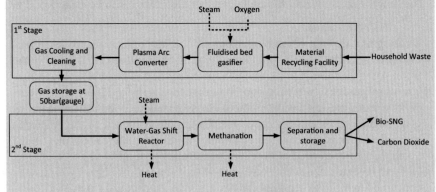

Fig. 5.5 Process diagram at the Swindon Bio-SNG plant

The Bio-SNG production process at Swindon consists of two main Stages. The 1st Stage is the thermochemical processing (gasification) of household waste to produce a volatile gas mixture. The 2nd Stage is its water-gas shift reaction and methanation to produce Bio-SNG.

The thermochemical process (the 1st Stage) is a combination of two steps. The first step is a gasification process in which steam and oxygen are used to partially oxidise the waste. In the second step, the gas emanating from the gasification process is exposed to high temperatures (around 1200 °C) in a plasma arc converter (A plasma torch powered by an electric arc is used to ionize gas and catalyse organic matter into a mixture of hydrogen, carbon monoxide and carbon dioxide with slag).

The gas mixture is then cooled to below 200 °C and treated to remove any residual particulates and gas contaminants which may include nitrogenous compounds, chloride, fluoride and sulphur gases.

The resulting gas mixture is stored in compressed gas storage vessels at 50 bar (gauge) pressure.

In the 2nd Stage the compressed gas is subject to water-gas shift and methanation reactions to transform the mixture of hydrogen, carbon monoxide and carbon dioxide from the gasification to a methane rich-gas and carbon dioxide (Fig. 5.6).

Fig. 5.6 Bio-SNG Pilot plant in Swindon, UK (Left-Gasification plant, Right—Water-gas shift and Methanation plant). With permission from Cadent, UK

The project has led to the construction of a commercial facility that will convert 10,000 tonnes of waste into 22 GWh of low carbon Bio-SNG each year. The Bio-SNG produced by the facility is to be injected to the Wales and West Utilities gas distribution network and to be used in a local compressed natural gas filling station.

For more information on the Go Green Gas project visit https://gogreengas. com.

5.3.3   *Hydrogen*

Hydrogen is the world's most abundant element and in its gaseous form is a high energy density fuel used in energy intensive applications such as welding and steel making. Hydrogen is a zero carbon-zero emission fuel and does not release any harmful by products when burnt. It is considered one of the cleanest fuels to be used in a low carbon energy system and is finding applications across the energy sector including transport (e.g. hydrogen vehicles) and producing electricity and heating (e.g. fuel cell CHP units).

The combustion of hydrogen is an exothermic oxidation process which produces thermal energy and pure water (H_2O) as a by-product.

$$2H_2(g) + O_2(g) \rightarrow 2H_2O(g) + 572 \text{ kJ of thermal energy}(286 \text{ kJ/mol of hydrogen})$$

where, H_2—hydrogen; O_2—oxygen; H_2O—water; (g)—gas

Hydrogen can also be used in an electrochemical cell (fuel cell) with oxygen or another oxidizing agent to convert the chemical energy into electricity without combustion. This is a more thermodynamically efficient process than producing electricity through the combustion of hydrogen.

Even though hydrogen is a clean fuel, fossil fuels are the dominant source of industrial hydrogen production. There are multiple methods for producing hydrogen, of which the following are the most popular:

- **Steam reforming of methane**: is the most cost effective method of producing hydrogen at large scale. This process requires heating natural gas to between 700 and 1100 °C in the presence of steam and a nickel catalyst. This breaks up the methane molecules to produce carbon monoxide and hydrogen. The carbon monoxide undergoes a water gas shift reaction to obtain further quantities of hydrogen. The steam reforming process produces carbon monoxide and carbon dioxide as by-products. Ongoing research investigates whether the carbon dioxide can be sequestrated in an oil or gas reservoir.
- **Electrolysis**: Electricity is used to split water into hydrogen and oxygen in an electrolyser. There is significant interest in the electrolysis of water using electricity from renewable sources (wind and solar electricity).
- **Coal/biomass gasification**: A fuel gas mixture containing primarily of hydrogen and carbon monoxide is produced by the gasification of both coal and biomass. This gas mixture then undergoes a water-gas shift reaction where carbon monoxide reacts with steam and increases the quantity of hydrogen. Similar to the steam reforming process of natural gas this route produces carbon monoxide and carbon dioxide as by-products.

The coal gas used in the early days of gas development contained over 50% (by volume) of hydrogen in the gas mixture delivered to consumers. Hydrogen is re-emerging as a suitable low carbon alternative to substitute natural gas in the gas grids. Several methods to substitute natural gas using hydrogen are being considered.

A. Convert existing gas networks to transport 100% hydrogen
B. Blending hydrogen with the natural gas at different quantities in the gas grid
C. Convert hydrogen to methane prior to gas grid injection.

A. Pure hydrogen transport

The concept of a pure hydrogen network has been proposed in many studies advocating energy systems that use hydrogen as the main energy vector for transport, electricity generation and heating. Several pioneering projects have commenced feasibility studies on the conversion of existing gas grids to transport 100% hydrogen (e.g. the H21 Leeds City Gate project in the UK [4]).

However, in the near-term a 100% hydrogen network is unlikely for the following reasons.

- Hydrogen is an expensive fuel due to its high cost of production. Excess renewable electricity (e.g. wind, solar plants) will not produce sufficient gas to replace current levels of natural gas demand.
- Gas transmission pipelines are mostly made of steel and hydrogen has an embrittling effect on steel.
- The current gas appliances are not compatible with hydrogen combustion; therefore a complete conversion of consumer gas appliances would be required.
- Safety concerns and public acceptance with regard to the transport and combustion of hydrogen.

Box 3: H21 Leeds City Gate Project

H21 Leeds City Gate is a project feasibility study to determine the technical and economic viability of converting the existing natural gas network in Leeds, one of the largest UK cities, to 100% hydrogen.

Hydrogen production and carbon capture and storage

The project proposed the use of steam methane reformers (SMR) with a hydrogen production capacity of 305,000 Nm^3/h (with 90% carbon dioxide capture) to serve the gas demand of Leeds. The captured carbon dioxide is to be compressed to a high pressure of 140 bar (gauge) and transported via a dedicated carbon dioxide transmission line to permanent sequestration in the North Sea. It is estimated that the project would sequester 1.5 million tonnes of carbon dioxide per year. Salt caverns in Teeside and the East Humber coast are to be used to provide intra-day and inter-seasonal storage of hydrogen and natural gas to supply the 1 in 20 peak demand and the 40 days of maximum average daily demand (coldest year) design requirements of the GB gas network.

Hydrogen Transmission System
A dedicated hydrogen transmission pipeline system was proposed to connect the SMR plant and salt cavern storage to Leeds. This pipeline should be capable of transporting the 1 in 20 peak day gas supply requirement of Leeds.

Medium Pressure and Low Pressure Gas Distribution networks
The study claims that the existing medium pressure and low pressure gas networks can be converted to 100% hydrogen with relatively minor upgrades. It was recommended that the existing gas network be segmented and converted from natural gas to hydrogen incrementally through the summer months over a three year period to minimize disruption to customers.

Appliance conversion
It was recognised that few models of hydrogen appliances and equipment for domestic, commercial and industrial sectors were available. It was recommended that a firm long-term plan and significant stimulus would be needed to provide the motivation to develop and produce the wide range of equipment required.

The H21 Leeds City Gate project estimated a net carbon dioxide saving for the area of conversion as 927,000 tonnes of carbon dioxide per year. It estimated a capital cost of £2.05 billion and an ongoing cost of £139 million each year for the project.

For more information on the H21 Leeds City Gate project visit https://www.northerngasnetworks.co.uk/wp-content/uploads/2017/04/H21-Report-Interactive-PDF-July-2016.compressed.pdf.

B.   Blending hydrogen in the gas grid

Blending small quantities of hydrogen with natural gas (up to 20% by volume) is proposed as a transitional step to reduce the carbon footprint of using natural gas. The gas industry is carrying out research and development to determine the level of hydrogen that could be used by consumers safely and with no changes to existing gas appliances.

A number of demonstration projects are piloting different schemes of hydrogen production and the blending of hydrogen in the gas grid. The quantity of hydrogen injection allowed is determined by regulations adopted in each individual country. For example in Germany a range of between 2 and 10% of hydrogen by volume (once mixed with natural gas) can be injected into the gas grid depending on the types of gas appliances that are connected downstream. If a compressed natural gas vehicle refuelling station is connected then the hydrogen injection must be reduced to 2% by volume due to its effect in vehicle fuel systems. Figure 5.7 shows an illustration of the limits of hydrogen allowed in the natural gas blend in different European countries.

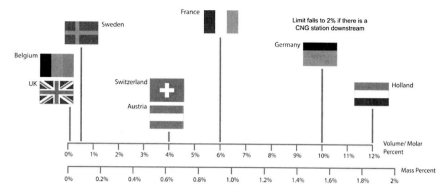

Fig. 5.7 Hydrogen injection limits in different European countries. Adapted from ITM Power report on Power-to-Gas: A UK feasibility study report (2013) [5]

The HyDeploy project in the UK is expected to trial blending up to 20% hydrogen into the natural gas supply [6].

> **Box4: HyDeploy Project at Keele University Campus**
>
> HyDeploy is a research project in the UK that has been given permission by the UK Health and Safety Executive to run a year long trial of blending hydrogen and natural gas on part of the private gas network at Keele University campus in Staffordshire. The project is delivered by a consortium of partners led by Cadent and Northern Gas Networks. It will be the first project in the UK to inject hydrogen into a natural gas network.
>
> The HyDeploy project seeks to demonstrate that a blend of hydrogen and natural gas can be used by gas customers without any changes to their behaviour or existing gas appliances. It aims to establish the level of hydrogen that can be safely blended with natural gas for use in a UK gas network.
>
> Extensive work was undertaken to ensure that all gas appliances operated safely with up to 20% hydrogen blend in natural gas (by volume). All households, in the trial area would undergo standard gas safety checks ahead of the trial to ensure the gas appliances and installations meet current safety standards.
>
> Hydrogen will be produced using an electrolyser fuelled by renewable electricity.
>
> The mixing of hydrogen and natural gas is to be done in a special injection and mixing unit which will be carefully controlled and monitored to ensure that the hydrogen and natural gas blend remains consistent as gas flows vary.
>
> *For more information on the HyDeploy project visit* https://hydeploy.co.uk/.

C. Methanation of hydrogen

Converting hydrogen to methane before injection into the gas grid has been proposed in order to avoid the implications of altering the physical and chemical properties of the gas mixture from an increased level of hydrogen. Methane is the primary component of natural gas and its injection does not change the characteristics of natural gas significantly. This would avoid the significant costs of converting the gas grid to operate on hydrogen.

Most of the research and development has been carried out on using the Sabatier thermochemical reaction for methanation of hydrogen and carbon dioxide in the presence of a catalyst.

The Sabatier thermochemical reaction can be expressed as,

$$CO_2(gas) + 4H_2(gas) \leftrightarrow CH_4(gas) + 2H_2O(gas)$$

Endothermic reaction requiring 252.9kJ/molof CO_2

where, CO_2—carbon dioxide; H_2—hydrogen; CH_4—methane; H_2O—water; (g)—gas

Methanation of hydrogen is an energy intensive process which is still undergoing research and development.

5.4   Challenges to Injecting Alternative Gases into the National Gas Grid

The regulatory, commercial and technical arrangements of the current gas system are based on the transport and utilisation of natural gas of a specified quality. For example, in the UK the content and characteristics of the gas to be transported in the national gas grid is specified by the Health and Safety Executive (HSE) and are legally enforced by the Gas Safety (Management) Regulations (1996). This document is addressed to gas transporters and other parties (gas shippers and terminal operators, network operators) and specifies that the gas shall be in accordance with the values specified in the Table 5.1.

However, the content and characteristics of the available substitute gases (described in Sect. 5.2–5.3) differ from that of natural gas (see Table 5.2).

A comprehensive assessment of the gas system needs to be undertaken to assess its compatibility to accommodate new gas sources without major implications. It is important to remember the expense and disruption of the system conversion that was required when the UK switched from coal gas to North Sea natural gas.

Extensive research has been carried out to assess the interchangeability of gas mixtures for gas appliances. Gas interchangeability refers to the ability of a gas appliance to perform satisfactorily and without readjustment on different fuel gases

Table 5.1   The content and characteristics of the gas transported in the UK gas grid

Content or characteristic	Value
Hydrogen sulphide content	≤ 5 mg/Nm^3
Total sulphur content (including Hydrogen sulphide)	≤ 50 mg/Nm^3
Hydrogen content	$\leq 0.1\%$ (molar)
Oxygen content	$\leq 0.2\%$ (molar)
Impurities	Shall not contain solid or liquid material which may interfere with the integrity or operation of pipes or any gas appliance which a consumer could reasonable be expected to operate
Hydrocarbon dew point and water dewpoint	Shall be at such levels that they do not interfere with the integrity or operation of pipes or any gas appliance which a consumer could reasonably be expected to operate
Wobbe Number (WN)	47.2 MJ/Nm$^3 \leq$ WN $\leq$ 51.41 MJ/Nm^3
Incomplete combustion factor	≤ 0.48
Soot index	≤ 0.6

The gas shall have been treated with a suitable stanching agent to ensure that it has a distinctive and characteristic odour which shall remain distinctive and characteristic when the gas is mixed with gas which has not been so treated, except that this paragraph do not apply where the pressure above ≥ 7 bar(gauge)

The gas shall be at a suitable pressure to ensure safe operation of any gas appliance which a consumer could reasonably be expected to operate

Adopted from A guide to the Gas Safety (Management) Regulations 1998 by Health and Safety Executive

that vary in their composition. The Wobbe Number is widely used as an indicator to assess the interchangeability of fuel gas mixtures.

The Wobbe Number is defined as,

$$\text{Wobbe Number} = \frac{\text{Higher Calorific Value of the gas mixture}}{\sqrt{\text{Specific Density of gas}}} \tag{5.1}$$

The Wobbe Number is proportional to the heat input to a gas appliance at constant pressure. It is used along with the flame speed factor as a first indication of the interchangeability of gas mixtures.

In addition to the performance of gas appliances, there are a number of other considerations that require careful investigation prior to the introduction of alternative gas sources into the gas grid. Some of these that would need to be investigated are the

Table 5.2 Content and characteristics of gases

Fuel	Typical composition							Calorific value (MJ/Nm^3)	Specific density	Wobbe index (MJ/Nm^3)
	CH_4	C_xH_y	H_2	H_2S	CO	CO^2	Other			
Coal gas	~35%		~50%		~10%			10–20	~0.58	22.5–30
Natural gas	~>90%	~3%					~2% N_2	~39	~0.6	48–53
Shale gas, tight gas	Similar to natural gas									
Coal bed methane	>95%	~3%				~2%		~37	~0.58	~53
Biogas	~60%		0–1%	0.1–0.5%		~25–50%	0–10% N_2; 0–0.5% O_2	17–25	~0.8	~19.5
Biomethane/coal-Bio-Synthetic Natural Gas	>95%					<0.1		~38	0.58	49
Hydrogen			100%					12.75	0.07	48

- **durability of all components of the gas network and the integrity of pipelines**. For example the performance of the steel based materials (e.g. their toughness and fatigue properties) can be reduced if hydrogen is in direct contact with the metal surface for extended periods. High-pressure gas pipelines in gas grids are typically of steel construction and will require detailed assessment if hydrogen is to be transported.

- **impact on condition monitoring procedures**. Condition monitoring of gas network equipment is an important aspect of asset management and gas network operation. Introducing new gas mixtures may require modifications to existing practices of condition monitoring in gas networks. For example, systems in use monitor corrosion defects in pipes. However, introducing hydrogen would increase the risks of cracks or crack-like defects in pipes and so would require different monitoring techniques to be adopted.

- **implication on safety of the transmission, distribution and use of the gas**. The design, construction, operation and maintenance procedures of the existing gas pipeline network are based on natural gas. The safety of the gas infrastructure and the risk posed to the public by the conveyance and use of natural gas in the gas grids are well understood and considered acceptable. However, injecting new gas sources in the gas grids would require reviewing established procedures to understand the potential risks to the public and gas network assets.

- **impact on the energy capacity of the gas supply system and gaseous losses**. Existing methods for the assessment of energy capacity of the gas system (to supply a particular gas demand) are based on transporting natural gas. The injection of an alternative gas source alters the physical and chemical properties of the gas mixture being transported and can affect the energy capacity and gaseous losses of the gas network. For example, studies have shown that blending gaseous hydrogen with natural gas would reduce the calorific value of the gas mixture and therefore require higher gas flow rates in order to deliver the same amount of energy. This would in turn affect the levels of linepack gas storage being maintained. Studies have shown that adding hydrogen to natural gas would increase gas losses as hydrogen can permeate through pipe walls and cracks quicker than natural gas.

- **commercial frameworks and gas billing methodologies**. The existing commercial frameworks and gas billing methodologies are based on delivering natural gas of a specified quality to all consumers. This is recognised as a barrier for introducing alternative sources of gas, particularly low carbon gases. For example, in the UK biogas needs to be cleaned and upgraded to match the quality of natural gas and this requires expensive gas purifying processes as well as the introduction of carbon based gases (propane) to meet the gas quality criteria. This processing is primarily in order to meet current billing standards and regulations. A review of the existing commercial frameworks and billing methodologies to accommodate a wide variety of gas types would unlock the potential for alternative gases to be introduced into the gas networks.

References

1. Jones R (1978) A history of gas production in Wales. Wales Gas Printing Centre, Wales
2. National Gas Museum: Two hundred years of energy, Gas industry timeline. Available at http://www.nationalgasmuseum.org.uk/gas-industry-chronology/. Accessed on 15 Aug 2018
3. Go Green Gas Project information available at https://gogreengas.com/. Accessed on 16 Aug 2018
4. H21 Leeds City Gate Project report available at https://www.northerngasnetworks.co.uk/wp-content/uploads/2017/04/H21-Report-Interactive-PDF-July-2016.compressed.pdf.   Accessed on 16 Aug 2018
5. ITM Power, 2013. Hydrogen Grid Inject Project recommendations report, TSB project reference 130815
6. HyDeploy Project information available at https://hydeploy.co.uk/. Accessed on 16 Aug 2018

Glossary

Normal temperature and pressure (NTP) is defined as 20 °C and 101.325 kPa.

Combined heat and power unit (CHP) is an energy conversion technology that generates electricity and useful thermal energy.

Relative Density is the density of a gas with reference to air at normal temperature and pressure (NTP).

Specific volume is the ratio of the volume of gas to its mass. It is the reciprocal of gas density.

Compressibility factor of a gas is an indication of its divergence from the behaviour of an ideal gas. It is defined as the ratio of the specific volume of a gas to the specific volume of an ideal gas at standard temperature and pressure. Its value varies with temperature.

Viscosity is a measure of a fluid's resistance to flow. It affects the pressure drop incurred when gas flows through a pipe. The viscosity of a gas increases with temperature and therefore the resistance to flow increases when gas is heated. This is opposite to the behaviour of a liquid when heated.

Calorific value is the heat generated by burning a unit of gas (either unit volume or unit mass) in air under standard conditions.

Higher heating value (HHV) is the heat produced by the complete combustion of a fuel. It is obtained when the products of combustion are condensed.

Lower heating value (LHV) is obtained by subtracting the latent heat of vaporisation of the water vapour formed by the combustion from the higher heating value.

© The Author(s), under exclusive licence to Springer Nature Switzerland AG 2020
M. Qadrdan et al., *The Future of Gas Networks*,
SpringerBriefs in Energy, https://doi.org/10.1007/978-3-319-66784-3

Printed in the United States
By Bookmasters